Acknowledgments

First edition: This manual was originally developed as part of a U.S. Forest Service technology transfer effort to get research findings off the shelf and into the hands of people who need them. A team of scientists and pest specialists worked with a writer-editor to summarize and compile more than 10 years of research on Christmas tree pests and their control. Much of this early work was done at what was then the North Central Forest Experiment Station, headquartered in St. Paul, MN. The first edition was published in 1983.

Thomas Nicholls, Louis Wilson, Darroll Skilling, and Marguerita Palmer were key contributors to the first edition from the North Central Forest Experiment Station. Kathryn Robbins and Peter Rush were contributors from the U.S. Forest Service, Northeastern Area State and Private Forestry. Janine Benyus edited the first edition. Much of their original text still remains in the newest edition.

Second Edition: In 1997 and 1998, the original manual was reviewed and updated. It was published in 1998 as Michigan State University Extension Bulletin E-2676 and edited by Deborah McCullough, Steven Katovich, Michael Ostry, and Jane Cummings Carlson.

Third Edition: The newest edition is published by the U.S. Forest Service, Northeastern Area State and Private Forestry. This edition includes a number of new pests, many of them found on fir Christmas trees. The original manual had a strong emphasis on Scotch pine. Most of the original pest descriptions remain in the manual; however, given that Christmas tree growers are growing more fir and other species, new pest problems needed to be addressed.

Editors included Steven Katovich, U.S. Forest Service, Northeastern Area State and Private Forestry; Deborah McCullough, Michigan State University; Michael Ostry, U.S. Forest Service, Northern Research Station; Jill O'Donnell, Michigan State University; Isabel Munck, U.S. Forest Service, Northeastern Area State and Private Forestry; and Cliff Sadof, Purdue University. Extensive contributions were provided by Ronald Kelly (retired), Vermont Department of Forests, Parks and Recreation; Thomas Nicholls (retired), U.S. Forest Service, North Central Forest Experiment Station; Jill Sidebottom, North Carolina State University; and Drew Carleton, Canadian Forest Service, Natural Resources Canada. Other contributors included Richard Cowles, The Connecticut Agricultural Experiment Station; Dennis Fulbright, Michigan State University; and Anette Phibbs and Christopher Deegan, Wisconsin Department of Agriculture, Trade and Consumer Protection. Numerous photographs were obtained from photographers individually and through the Bugwood.org Web site.

The editors are very grateful to Juliette Watts and Sandy Clark, U.S. Forest Service, Northeastern Area State and Private Forestry, for assistance in publication design and layout, editing, and a myriad of other items.

Page

Index A
Common and Scientific Names of
Pests and Trees...................................5

Index B
Potential Pest Problems By Tree Species..........8

Introduction 11
How to Use This Manual 11
What is a Pest?.................................... 12
Insects and Mites 12
Fungi and Nematodes 12
Birds and Mammals 13
Environmental Factors........................... 13
Symptoms and Signs of Tree Injury.................. 14
Discolored Foliage 14
Missing Foliage 15
Deformed and Stunted Tissue 15
Pitch Flow... 15
Wood Shavings 15
Insect and Pathogen Parts and Structures 16
Root Injury 16
Pest Management................................. 16
Step 1—Plant the Right Species on the Right
Site and Invest in Quality Planting Stock 17
Choosing Your Trees 17
Preparing the Planting Site 17
Planting.. 17
Step 2—Monitor and Scout Trees for Pests..... 18
Step 3—Use a Combination of Controls 19
Manual and Mechanical Control 19
Biological Control 19
Cultural Control 20
Chemical Control................................... 20
Step 4—Evaluate Your Control Efforts 21
**Using Degree Day Accumulation for Improving
Timing of Insect Pest Management**............ 22
Where to Get Help................................ 25
How to Submit Materials for Identification......25
Digital Images 25

Page

Needle Discoloration or Distortion.................. 27
Admes Mite 28
Air Pollution Injury 29
Balsam Gall Midge 30
Balsam Twig Aphid................................ 32
Brown Spot Needle Blight 34
Cyclaneusma Needlecast 35
Dothistroma Needle Blight 36
Douglas-Fir Needle Midge 37
Drought Injury....................................... 38
Elongate Hemlock Scale 39
Eriophyid Mites..................................... 40
Fall Needle Drop.................................... 41
Fir Needle Rust...................................... 42
Herbicide Injury 43
Lirula and Isthmiella Needlecast.............. 44
Lophodermium Needlecast...................... 46
Naemacyclus Needlecast
(see Cyclaneusma Needlecast)
Pine Needle Rust 48
Pine Needle Scale 50
Pine Thrips 51
Ploioderma Needlecast........................... 52
Rhabdocline Needlecast.......................... 53
Rhizosphaera Needle Blight of Firs 54
Rhizosphaera Needlecast of Spruce............... 55
Salt Injury 57
Spruce Needle Rusts 58
Spruce Spider Mite 59
Swiss Needlecast................................... 60
Winter Injury 61
Needle Feeding 63
Bagworm.. 65
Balsam Fir Sawfly.................................. 66
European Pine Sawfly 67
Grasshoppers.. 69
Gypsy Moth 70
Introduced Pine Sawfly 72

Page

Jack Pine Budworm 73

Northern Conifer Tussock Moth
(see Pine Tussock Moth)

Pine Chafer (Anomala Beetle)...........................74

Pine False Webworm.................................75

Pine Needle Midge 76

Pine Tube Moth 77

Pine Tussock Moth 78

Pine Webworm 79

Redheaded Pine Sawfly 80

Spruce Budworm 82

Spruce Fir Looper 83

Spruce Needleminers...............................84

Shoot/Branch Injury 87

Aphids 89

Balsam Shootboring Sawfly.........................90

Balsam Woolly Adelgid.............................91

Broom Rust of Fir 93

Cytospora Canker (see Leucostoma Canker)

Deer..94

Delphinella Shoot Blight 95

Diplodia Shoot Blight and Canker 96

Eastern Pine Shoot Borer...........................98

Eastern Pine Weevil 100

European Pine Shoot Moth 101

Frost Injury......................................102

Gremmeniella Canker (see Scleroderris Canker)

Jack Pine Tip Beetle..............................103

Leucostoma Canker 104

Nantucket Pine Tip Moth..........................105

Pales Weevil 106

Phomopsis Canker 108

Pine Grosbeak...................................109

Pine Root Tip Weevil 110

Pine Shoot Beetle................................ 111

Pine Spittlebug 113

Pine Tortoise Scale............................... 114

Saratoga Spittlebug 115

Page

Scleroderris Canker.......................... 117

Sirococcus Shoot Blight...................... 118

Sphaeropsis Shoot Blight and Canker
(see Diplodia Shoot Blight and Canker)

Spruce Bud Scale 119

White Pine Blister Rust...................... 120

White Pine Weevil........................... 122

Shoot/Branch Galls........................ 125

Cedar-Apple Rust 127

Cooley Spruce Gall Adelgid 128

Eastern Gall Rust............................ 130

Eastern Spruce Gall Adelgid 131

Gall Rusts (see Eastern Gall Rust)

Northern Pitch Twig Moth 132

Spruce Gall Midge........................... 133

Western Gall Rust (see Eastern Gall Rust)

Dead Tree and Stem/Root Injury..................... 135

Allegheny Mound Ant 137

Armillaria Root Rot 138

Conifer Root Aphid.......................... 139

Gopher (see Pocket Gopher or Thirteen-Lined
Ground Squirrel)

Meadow Vole and Pine Vole.................. 149

Mouse (see Meadow Vole and Pine Vole)

Phytophthora Root Rot 141

Pine Bark Adelgid........................... 142

Pine Root Collar Weevil 143

Pinewood Nematode......................... 145

Pocket Gopher.............................. 146

Rabbit and Snowshoe Hare 147

Thirteen-Lined Ground Squirrel 148

Voles (see Meadow Vole and Pine Vole)

White Grubs 151

Wood Borers and Bark Beetles............... 153

Yellow-Bellied Sapsucker.................... 155

Zimmerman Pine Moth....................... 156

General Index 158

Common Name	Scientific Name	Page

Pests

Admes mite	*Eurytetranycus admes*	28
Allegheny mound ant	*Formica exsectoides*	137
Anomala beetle—see pine chafer		
aphids (also see balsam twig and conifer root aphids)	*Cinara* spp. and *Eulachnus* spp.	89
Armillaria root rot	*Armillaria* spp.	138
bagworm	*Thyridopteryx ephemeraeformis*	65
balsam fir sawfly	*Neodiprion abieties*	66
balsam gall midge	*Paradiplosis tumifex*	30
balsam shootboring sawfly	*Pleroneura brunneicornis*	90
balsam twig aphid	*Mindarus abietis*	32
balsam woolly adelgid	*Adelges piceae*	91
broom rust of fir	*Melampsorella caryophyllacearum*	93
brown spot needle blight	*Mycosphaerella dearnessii*	34
cedar-apple rust	*Gymnosporangium juniperi-virginianae*	127
conifer root aphid	*Prociphilus americanus*	139
Cooley spruce gall adelgid	*Adelges cooleyi*	128
Cyclaneusma needlecast (= Naemacyclus)	*Cyclaneusma minus*	35
Cytospora canker—see Leucostoma canker		
deer	*Odocoileus virginianus*	94
Delphinella shoot blight	*Delphinella balsameae*	95
Diplodia shoot blight and canker	*Diplodia pinea*	96
Dothistroma needle blight	*Mycosphaerella pini*	36
Douglas-fir needle midge	*Contarinia pseudotsuga*	37
eastern gall rust	*Cronartium quercuum*	130
eastern pine shoot borer	*Eucosma gloriola*	98
eastern pine weevil	*Pissodes nemorensis*	100
eastern spruce gall adelgid	*Adelges abietis*	131
elongate hemlock scale	*Fiorinia externa*	39
eriophyid mites	*Setoptus* spp. and *Nalepella* spp.	40
European pine sawfly	*Neodiprion sertifer*	67
European pine shoot moth	*Rhyacionia buoliana*	101
fir needle rust	*Uredinopsis* spp. and *Milesina* spp.	42
grasshoppers	*Melanoplus* spp.	69
green spruce needleminer	*Epinotia nanana*	84
Gremmeniella canker—see Scleroderris canker		
gypsy moth	*Lymantria dispar*	70
hare, snowshoe	*Lepus americanus*	147
introduced pine sawfly	*Diprion similis*	72
Isthmiella needlecast	*Isthmiella faullii*	44
jack pine budworm	*Choristoneura pinus pinus*	73
jack pine tip beetle	*Conopthorus resinosae*	103

Common Name	Scientific Name	Page
Leucostoma canker	*Leucostoma kunzei*	104
Lirula needlecast	*Lirula nervata, Lirula mirabilis*	44
Lophodermium needlecast	*Lophodermium seditiosum*	46
meadow vole (= meadow mouse)	*Microtus pennsylvanicus*	149
Naemacyclus needlecast—see Cyclaneusma needlecast		
Nantucket pine tip moth	*Rhyacionia frustrana*	105
northern conifer tussock moth	*Dasychira plagiata*	78
northern pitch twig moth	*Retinia albicapitana*	132
Pales weevil	*Hylobius pales*	106
Phomopsis canker	*Phomopsis* spp.	108
Phytophthora root rot	*Phytophthora cinnamomi; Phytophthora* spp.	141
pine bark adelgid	*Pineus strobi*	142
pine bark beetle (= pine engraver)	*Ips* spp.	153
pine chafer (Anomala beetle)	*Anomala oblivia*	74
pine false webworm	*Acantholyda erythrocephala*	75
pine grosbeak	*Pinicola enucleator*	109
pine needle midge	*Contarinia baeri*	76
pine needle rust	*Coleosporium asterum*	48
pine needle scale	*Chionaspis pinifoliae*	50
pine root collar weevil	*Hylobius radicis*	143
pine root tip weevil	*Hylobius rhizophagus*	110
pine shoot beetle	*Tomicus piniperda*	111
pine spittlebug	*Aphrophora parallela*	113
pine thrips	*Gnophothrips* spp.	51
pine tortoise scale	*Toumeyella parvicornis*	114
pine tube moth	*Argyrotaenia pinatubana*	77
pine tussock moth	*Dasychira pinicola*	78
pine vole	*Microtus pinetorum*	149
pine webworm	*Pococera robustella*	79
pinewood nematode	*Bursaphelenchus xylophilus*	145
pitch nodule maker—see northern pitch twig moth		
Ploioderma needlecast	*Ploioderma lethale*	52
pocket gopher	*Geomys bursarius*	146
rabbit, cottontail	*Sylvilagus floridanus*	147
redheaded pine sawfly	*Neodiprion lecontei*	80
Rhabdocline needlecast	*Rhabdocline pseudotsugae*	53
Rhizosphaera needle blight of firs	*Rhizosphaera pini*	54
Rhizosphaera needlecast of spruce	*Rhizosphaera kalkhoffii*	55
Saratoga spittlebug	*Aphrophora saratogensis*	115
Scleroderris canker (= Gremmeniella canker)	*Gremmeniella abietina*	117
Sirococcus shoot blight	*Sirococcus conigenus*	118
Sphaeropsis shoot blight and canker—see Diplodia shoot blight and canker		

Common Name	Scientific Name	Page
spruce bud scale	*Physokermes piceae*	119
spruce budworm	*Choristoneura fumiferana*	82
spruce fir looper	*Macaria signaria*	83
spruce gall midge	*Mayetiola piceae*	133
spruce needle rusts	*Chrysomyxa* spp.	58
spruce needleminers	*Taniva albolineana*, *Epinotia nanana*, and *Coletechnites piceaella*	84
spruce spider mite	*Oligonychus ununguis*	59
Swiss needlecast	*Phaeocryptopus gäeumannii*	60
thirteen-lined ground squirrel	*Spermophilus tridecemlineatus*	148
webworms—see pine false webworm and pine webworm		
western gall rust	*Peridermium harknessii*	130
white grubs	*Phyllophaga* spp., *Polyphylla* spp., and *Rhizotrogus majalis*	151
white pine blister rust	*Cronartium ribicola*	120
white pine weevil	*Pissodes strobi*	122
wood borers and bark beetles	*Monochamus* spp., etc.; *Ips* spp.	153
yellow-bellied sapsucker	*Sphyrapicus varius*	155
Zimmerman pine moth	*Dioryctria zimmermani*	156

Trees

Austrian pine	*Pinus nigra*	8
balsam fir	*Abies balsamea*	8
Black Hills spruce	*Picea glauca* var. *densata*	10
Colorado blue spruce	*Picea pungens*	9
Douglas-fir	*Pseudotsuga menziesii*	8
eastern redcedar	*Juniperus virginiana*	8
eastern white pine	*Pinus strobus*	9
Fraser fir	*Abies fraseri*	8
Norway spruce	*Picea abies*	9
red pine	*Pinus resinosa*	9
Scotch pine (also called Scots pine)	*Pinus sylvestris*	9
white fir (also called concolor fir)	*Abies concolor*	8
white spruce	*Picea glauca*	10

Tree Species   Pests – including environmental factors

All Trees (The following can impact most tree species though some species may be more susceptible than others.)

Air pollution, deer, drought injury, fall needle drop, frost injury, herbicide injury, meadow vole and pine vole, pocket gopher, rabbit and hare, salt injury, thirteen-lined ground squirrel, winter injury, yellow-bellied sapsucker

Douglas-fir (Check under "All Trees" for additional damaging agents)

Insects/mites: Allegheny mound ant, aphids, bagworm, bark beetles, Cooley spruce gall adelgid, Douglas-fir needle midge, eastern pine shoot borer, elongate hemlock scale, grasshoppers, gypsy moth, Pales weevil, pine needle scale, spruce budworm, spruce fir looper, spruce spider mite, white grubs, white pine weevil, wood borers

Diseases: Armillaria root rot, Rhabdocline needlecast, Phytophthora root rot, Scleroderris canker, Swiss needlecast

Eastern Redcedar (Check under "All Trees" for additional damaging agents)

Insects/mites: Allegheny mound ant, aphids, bagworm, bark beetles, grasshoppers, gypsy moth, juniper webworm (not in this manual), pine needle scale, spruce spider mites, white grubs, wood borers

Diseases: Armillaria root rot, cedar-apple rust

All Firs: *balsam, Fraser, white* (Check under "All Trees" for additional damaging agents)

Insects/mites: Allegheny mound ant, aphids, bagworm, balsam fir sawfly, balsam gall midge, balsam shootboring sawfly, balsam twig aphid, balsam woolly adelgid, conifer root aphid, elongate hemlock scale, eriophyid mites, grasshoppers, gypsy moth, Pales weevil, pine needle scale, pine spittlebug, pine tussock moth, Saratoga spittlebug, spruce fir looper, spruce spider mite, spruce budworm, white grubs, wood borers

Diseases: Armillaria root rot, fir needle rust, broom rust of fir, Delphinella shoot blight, Isthmiella needlecast, Lirula needlecasts, Phytophthora root rot, Rhizosphaera needle blight of firs, Scleroderris canker

All Pines: *Austrian, eastern white, red, and Scotch* (Check under "All Trees" for additional damaging agents)

Insects/mites: Allegheny mound ant, aphids, bark beetles, eastern pine shoot borer, eastern pine weevil, elongate hemlock scale, eriophyid mites, grasshoppers, gypsy moth, introduced pine sawfly, mites, jack pine budworm, northern pitch twig moth, Pales weevil, pine chafer, pine false webworm, pine needle scale, pine root collar weevil, pine shoot beetle, pine webworm, spruce spider mite, white grubs, white pine weevil, wood borers, Zimmerman pine moth

Diseases: Armillaria root rot, brown spot needle blight, Lophodermium needlecast, Phytophthora root rot, pinewood nematode, Scleroderris canker

Austrian Pine (Check under "All Trees" and "All Pines" for additional damaging agents)

Insects/mites: European pine sawfly, European pine shoot moth, Nantucket pine tip moth, pine bark adelgid, pine spittlebug, pine thrips, pine tortoise scale

Diseases: Diplodia shoot blight and canker, Dothistroma needle blight, Ploioderma needlecast

Tree Species   Pests – including environmental factors

Eastern White Pine   (Check under "All Trees" and "All Pines" for additional damaging agents)

Birds: Pine grosbeak

Insects/mites: bagworm, pine bark adelgid, pine root tip weevil, pine spittlebug, pine tube moth, pine tussock moth, Saratoga spittlebug, spruce fir looper

Diseases: white pine blister rust, very sensitive to air pollution

Red Pine   (Check under "All Trees" and "All Pines" for additional damaging agents)

Birds: Pine grosbeak

Insects/mites: European pine sawfly, European pine shoot moth, jack pine tip beetle, Nantucket pine tip moth, pine needle midge, pine root tip weevil, pine tortoise scale, pine tussock moth, redheaded pine sawfly, Saratoga spittlebug

Diseases: Diplodia shoot blight and canker, Dothistroma needle blight, pine needle rust, Ploioderma needlecast, Sirococcus shoot blight

Scotch Pine   (Check under "All Trees" and "All Pines" for additional damaging agents)

Birds: Pine grosbeak

Insects/mites: bagworm, European pine sawfly, European pine shoot moth, jack pine tip beetle, Nantucket pine tip moth, northern pitch twig moth, pine bark adelgid, pine needle midge, pine root tip weevil, pine spittlebug, pine thrips, pine tortoise scale, redheaded pine sawfly, Saratoga spittlebug

Diseases: Cyclaneusma (=Naemacyclus) needlecast, Diplodia shoot blight and canker, eastern gall rust, pine needle rust, Ploioderma needlecast, Sirococcus shoot blight, western gall rust

All Spruce: *Colorado blue, Norway, white, and Black Hills*   (Check under "All Trees" for additional damaging agents)

Birds: Pine grosbeak

Insects/mites: Admes mite, Allegheny mound ants, aphids, bagworm, bark beetles, eastern pine weevil, elongate hemlock scale, eriophyid mites, grasshoppers, gypsy moth, Pales weevil, pine grosbeak, pine needle scale, pine spittlebug, pine tussock moth, spruce bud scale, spruce budworm, spruce fir looper, spruce needleminers, spruce spider mites, white grubs, white pine weevil, wood borers

Diseases: Armillaria root rot, Leucostoma (=Cytospora) canker, Phomopsis canker, Phytophthora root rot, Scleroderris canker, spruce needle rusts

Colorado Blue Spruce   (Check under "All Trees" and "All Spruce" for additional damaging agents)

Insects/mites: Cooley spruce gall adelgid

Diseases: Rhizosphaera needlecast of spruces, Sirococcus shoot blight

Tree Species Pests – including environmental factors

White and Black Hills Spruce (Check under "All Trees" and "All Spruce" for additional damaging agents)

Insects/mites: Balsam fir sawfly, Eastern spruce gall adelgid, eastern pine shoot borer, spruce gall midge, yellow-headed spruce sawfly (not in this manual)

Diseases: Rhizosphaera needlecast of spruces, Sirococcus shoot blight

Introduction

This manual was designed to help you identify and control damaging pests of Christmas trees in the North Central and Northeastern regions of the United States. Most of the information provided also applies to the States and Canadian Provinces that border these regions.

You do not have to be a pest specialist to use this information. The manual was written in everyday language so that anyone with an interest in Christmas trees can read and understand it. Because it is meant to be a tool and not a textbook, we included only the basic information that you need to know to diagnose and manage pest problems in your plantations.

In addition to the illustrated pest profiles, you will find practical advice on:

- How to look for and recognize potential pests

- How to select, plant, and care for trees so they are less likely to be damaged by pests

- How to keep pests at low levels to prevent tree damage

The management techniques we provide will discourage pests and prevent them from causing serious damage. We encourage you to read the introductory section of the manual to become familiar with the kinds of potential pests that can affect your trees.

How to Use This Manual

Carry the manual with you when you inspect your nursery or plantation. If you notice anything out of the ordinary, determine the tree species that is affected and then simply follow six steps:

Needle Discoloration or Distortion

Needle Feeding

Shoot/Branch Injury

Shoot/Branch Galls

Dead Tree and Stem/Root Injury

1. Decide what kind of injury symptom your tree has.
Turn to the appropriate section in the manual. Check your selection by comparing the symptoms on your tree with the description that appears on the first page of each section.

2. Leaf through that section. Note the tree "Host" or "Species Affected" for each individual pest problem. This information will be located at the beginning of each individual pest writeup. List the pest problems that may impact the tree species you are growing.

3. Now, sort through the pests that can affect your species and find the photos of symptoms that most closely match the damage you see on your trees. When you are pretty sure you have identified the culprit, double check it against the **"Pests that cause similar symptoms"** list if provided.

4. Review the symptoms and signs listed under "Look For."
You can identify most pests by the clues they leave behind or by the kind of injury they cause. These symptoms and signs are highlighted in italics and grouped by the time of year they are most likely to be seen (timing may vary with geographic location). Features visible year round are listed first, without a calendar heading. Pests are also described in terms of their importance, biology, and other characteristics. If you have any doubts about identification, send samples of the pest and the injured tree parts to your local pest specialist (see "How to Submit Materials for Identification").

5. Decide whether control is needed.
The "Monitoring and Control" section for each pest can help you evaluate pest activity on your trees and help you decide if control measures are needed. Before you begin any pest control treatment, ask yourself whether the value of the benefits will exceed the cost of the treatment. In short, will it pay? You may want to contact a pest specialist to help you predict and estimate damage (see "Where to Get Help").

6. Select control methods.
The "Monitoring and Control" section for each pest provides management and control options for trees currently growing in your plantation. This section may also provide suggestions to help prevent or reduce pest problems on trees or established seedlings. "Next Crop" controls can help you guard against future pest problems the next time you plant.

What is a Pest?

A pest is an agent that has the potential to injure or damage trees, which reduces their value for their intended use. Insects, diseases, animals, birds, and environmental factors that destroy or damage those trees are therefore considered pests. In their natural settings, these "pests" may be relatively harmless or perhaps even beneficial. In intensively managed nurseries and plantations, however, they can be undesirable and may require prevention or control. The pests in this manual fall into four groups:

- Insects
- Fungi and nematodes
- Birds and mammals
- Environmental factors

The more you know about them, the better able you are to solve pest problems in your nursery or plantation. The following briefly describes how these pests grow and reproduce, and under what conditions they affect Christmas trees.

Insects and Mites

Insects and their close relatives, mites, are the most common pests of Christmas trees. This manual lists many common insect pests, but there are others that may cause injury to your trees.

When abundant, insects can cause costly injury at various times in a tree's growing cycle. Seedlings and young trees are particularly vulnerable because it may take only a few insects to injure or kill them. However, older trees may also be injured when insects are numerous.

Insects damage Christmas trees in many ways. They can chew on or inside the needles or tunnel inside the shoots and trunk. Some insects suck sap from the needles, buds, or stems, which may weaken or kill the tree. Others cause swellings, or galls, to form, and a few spread pathogens directly or indirectly while feeding.

Insects are one of those unique groups of organisms that change form at least once during their lifetime. This is important to you as a Christmas tree grower because different insect forms cause different kinds of damage. For example, the larval form of one species may cause serious injury to a certain tree species while the adult form is harmless.

The simplest kind of change, or metamorphosis, involves three life stages: egg, nymph, and adult. The immature stage, called

Redheaded pine sawfly larvae feeding on pine needles. (L.L. Hyche, Auburn Univ., Bugwood.org)

a nymph, is a miniature version of the adult insect. A nymph hatches from an egg and then molts, or sheds its skin, several times as it grows. Eventually the nymph molts one last time into a mature adult. Pests that have this simple metamorphosis include grasshoppers, thrips, spittlebugs, aphids, and mites.

More complex, or complete, metamorphosis occurs for insects that have four life stages: egg, larva, pupa, and adult. Like nymphs, larvae must molt several times. Larvae enter a pupal stage before emerging as an adult. Each stage is greatly different in form from previous stages. Larvae look much different than the adult. Common insects with complete metamorphosis are beetles, moths and butterflies, midges (flies), and sawflies.

In both simple and complete metamorphosis, the adults mate, produce eggs, and start the life cycle process over again. Most Christmas tree insect pests have one generation a year. Some insects and mites, however, may have two or more generations each year.

Fungi and Nematodes

Fungi and nematodes can cause disease in trees. Fungi cause the majority of diseases of Christmas trees. A diseased tree infected with fungi may have a wide range of symptoms, such as abnormal swellings on the branches, discolored needles, dropping of needles, pustules or blisters on the foliage, curling of the growing shoots, and cankers on branches or trunks. A canker is an area of dead tissue.

Rabbits can clip and kill seedlings. (T. Spivey, USFS, Bugwood.org)

Spore-filled fruitbodies erupt from needles infected with pine needle rust. (USFS - NCRS Archive, Bugwood.org)

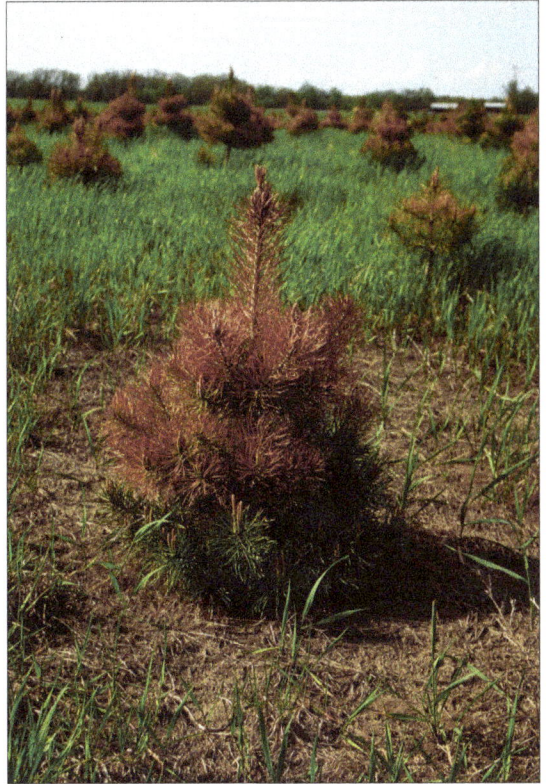

Winter burn injury on a young Scotch pine. (S. Katovich, USFS, Bugwood.org)

Fungi reproduce by means of spores—the fungal equivalent of seeds. Spores are produced in a wide variety of fungal "fruitbodies" that develop in bark, wood, or needles. Some fungal spores are windblown and can spread quickly; other spores are carried in rain water and move only as far as raindrops are splashed or blown. Spores can also be transported on equipment, such as shearing knives, and on infected plant material.

Nematodes are members of a group of animals known as roundworms—long, worm-like animals tapered at both ends. They are microscopic. Nematodes feed by puncturing tree cells with their hollow feeding tube and sucking out the cell contents. They hatch from eggs and pass through several larval stages when developing. Some nematodes that injure Christmas trees are carried from tree to tree by insects.

Birds and Mammals

Some birds and mammals can also injure Christmas trees. Pine grosbeaks eat buds and yellow-bellied sapsuckers peck holes in tree stems. Although birds usually cause only minor problems, they can sometimes cause enough injury to degrade trees.

Deer nip shoots and seedlings, occasionally causing extensive damage. Voles, rabbits, and gophers will chew the bark of stems or roots and can readily kill trees. Losses can sometimes be severe.

Environmental Factors

A variety of environmental factors can injure Christmas trees either directly or indirectly, including extreme weather and toxic chemicals, such as air pollutants, pesticides, salt, and excessive amounts of fertilizer. Disorders caused by these factors cannot spread from one tree to another like fungi, but they often weaken and predispose trees to other pests.

Symptoms and Signs of Tree Injury

When you are protecting your investment from very small insects, microscopic pathogens, or pests that feed underground, it helps to have an eye for detail. If you know what to look for when inspecting your trees, you can spot a pest problem in its early stages and greatly reduce losses and control costs.

Most likely, you will see the results of pest activity long before you notice the actual pest. An injured tree will have symptoms such as unusual color, missing foliage, and deformed parts, among others. Although these clues may help you diagnose the injury, they may also mislead you. For example, a symptom of one pest may look much like the injury caused by several other pests. To complicate matters further, two or more pests may injure the tree at the same time, producing a new symptom by their interaction. Symptoms also change; yellow foliage may redden, turn brown, or fall off entirely. Therefore, you cannot rely on symptoms alone when you diagnose tree injury.

Reading the signs of the pest as well as symptoms of the host is usually the best way to tell one type of injury from another. Signs are the physical evidence of insect and disease activity and include the pests themselves (eggs, larvae, fruitbodies); their enclosures (webs, cases, cocoons); debris (cast skins, wood slivers, pellets of waste); pitch flow; and associated insects or diseases, such as ants and sooty mold. Sometimes, two different pests will produce look-alike symptoms and leave the same signs. In that case, you or a pest specialist must examine and identify the pest itself (see "How to Submit Materials for Identification").

The detective work involved in identification becomes more difficult as time passes because both symptoms and signs change. A vigorous tree may mask or outgrow the injury. On the other hand, a weak tree will become susceptible to invading insects and pathogens that can confuse the diagnosis by producing symptoms and signs of their own. Signs such as webs, waste, or cast skins will also break down with time.

It is ideal to observe a symptom or sign during the early stages of pest activity. To catch pests in action, start monitoring or scouting your trees at planting time and continue to examine them frequently throughout their lives. The following section describes the major symptoms and signs to watch for every time you inspect your nursery or plantation.

Discolored Foliage

Discolored foliage can result from damage to any part of the tree—roots, trunk, branches, or needles. When a group of needles of the same age is discolored, damage is usually centered in the individual needles. For instance, needlecast fungi tend to infect needles of the same age and cause banding on the individual needles they infect. However, if only one branch is discolored, including needles of different ages, then the injury is usually within or at the base of

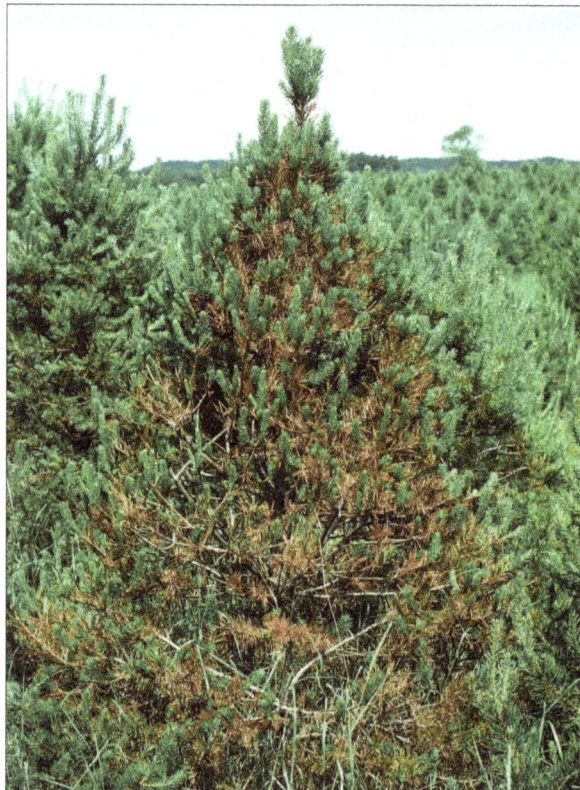

Needlecast diseases result in discolored needles that often drop. (USFS - NCRS Archive, Bugwood.org)

the branch. When the whole tree is evenly discolored, the injury is usually on the main stem or roots, or at the root collar.

Blackened foliage and/or bark indicate soft scales, aphids, or spittlebugs. A black, sooty mold grows on the sugary "honeydew" and spittle produced by these insects. Needles and shoots covered by the eggs or bodies of insects may also look discolored. In these instances, you can directly identify the insects injuring your trees.

If discolored foliage is only on one side of a tree, you can suspect an abiotic agent such as winter injury, salt, or herbicide.

Balsam gall midge injury creates deformed needles. (R.S. Kelley, VT Dept. of Forests, Parks and Recreation, Bugwood.org)

Sawfly larval feeding removes needles. (S. Katovich, USFS, Bugwood.org)

Missing Foliage

Needle loss is a common symptom of many insect and disease injuries. Foliage affected by winter injury or needlecast fungi will die and drop off early. Notched, broken, or hollowed-out needles indicate insect feeding. Insect foliage feeders will strip off clusters of needles, often in a characteristic pattern. Look for the insect or its frass, webs, cocoons, or cast skins on the surrounding foliage and beneath the injury; these are all signs of insect feeding (see "Insect and Pathogen Parts and Structures").

Deformed and Stunted Tissue

Insects and diseases can cause galls, swellings, and other kinds of abnormal growth on needles, shoots, stems, or roots. Past injuries from insects, diseases, or animals may cause excessive branching, forking, and crooking. Stunted shoots are caused by drought and frost damage, insects feeding on shoots and roots, or infection by shoot-blight fungi. Once weakened by injury or stress, trees often grow slowly. Although economically important, this growth loss may be difficult to detect and diagnose.

Pitch Flow

When insects feed or tunnel in the shoots, branches, and stems of living conifers, a pitchy substance commonly surrounds or flows from the point where

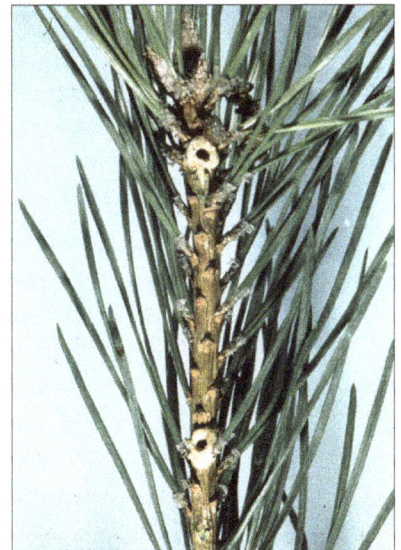

Pitch tubes formed at the entry hole of the pine shoot beetle. (E.R. Hoebeke, Cornell Univ., Bugwood.org)

they entered the tree. Sometimes this material creates popcorn-like balls referred to as a pitch tube. Canker and shoot-blight diseases may also cause pitch to flow from damaged areas.

Wood Shavings

Insects tunneling in stems and branches often produce fine sawdust or coarse slivers of wood. You may see piles of this material on the ground or stuck to the bark.

Insect and Pathogen Parts and Structures

Insects and pathogens often leave behind evidence. You can find cast skins of nymphs, larvae, or pupae, as well as old eggs that have hatched. You can often use this evidence to identify the species of insect. Several species of needle feeders construct protective bags or webs of silks, or cocoons that are distinctive. Some wood-boring insects make pupal cells lined with wood fibers, called chip cocoons, in the wood where they feed.

Small pellets of frass (insect waste) left by foliage-feeding insects can often be found near or beneath damaged foliage. Spittlebug nymphs produce white, frothy masses resembling spittle on the twigs or branches of trees. Part the mass carefully to see the small, soft nymph.

Fungi produce spores in reproductive structures called fruitbodies. These small structures are formed in the dead tissues of needles, shoots, and stems. Some species of fungi

Digging up a tree can reveal root injury. (R. Cowles, CT Agric. Exp. Stn.)

form characteristic sheets of fungal material called mycelial fans (see "Armillaria root rot"). Keep in mind that fungi found in dead tissues may or may not have killed the tree. Many fungi are secondary agents, which means they invade tissues after the tree has been injured or killed by something else.

Root Injury

Sometimes, damage to trees occurs below ground to the roots or root collar of a tree. A shovel can be a very useful tool when diagnosing below-ground pest problems. White grubs, root collar weevils, and root rots such as Armillaria root disease attack below ground and can cause damage that kills trees outright. Look for pitch or resin leaking into the soil, sometimes forming black, sticky masses of soil. Fine roots might be missing or larger roots may be hollowed out or covered with fungal mats. In some cases you may find insects or mushrooms. Excavating

Some weevil species form chip cocoons. (C. Evans, IL Wildlife Action Plan, Bugwood.org)

roots can also uncover planting issues such as J-rooting that makes trees more prone to pest problems.

Pest Management

Begin pest management before you plant your first tree and extend it through harvest time. In fact, everything you do to your plantation—selecting a species, preparing the site, planting, shearing, and harvesting—can influence pest activity. Some early planning and a strategy that focuses on pest prevention can be worthwhile.

If a pest becomes a problem, reduce economic losses to a tolerable level by using a variety of control methods. Emphasize cultural and biological treatments, supplemented only when necessary by pesticides. Using all available pest control strategies in a complementary way is called IPM—Integrated Pest Management. Four general steps are involved in IPM.

Step 1. Plant the Right Species on the Right Site and Invest in Quality Planting Stock

Choosing Your Trees

Ideally, the species you plant should grow well on your site and be somewhat resistant to pests. From a health standpoint, other important traits might include hardiness and tolerance to heat, cold, or drought. Be very aware of the types of soils you have. Are they nutrient rich or poor? Are they well or poorly drained? As a general rule, sandy, well-drained soils low in nutrients will support pines better than firs or spruces. Trees that are stressed and growing slowly are often more likely to be successfully attacked by insects or pathogens. Because Scotch pine was a favored Christmas tree species for many years, several varieties have been developed to enhance certain genetic qualities. Some of these qualities relate to pest resistance. So if you are planning to plant Scotch pine, check with local growers and Extension specialists for the pests most common to your area and select a variety of Scotch pine that is most resistant to those pests.

Healthy, vigorous seedlings can develop into young trees that are better able to withstand pest problems. Make sure seedlings are free of pest problems when they arrive from the nursery.

Preparing the Planting Site

Careful site preparation is another essential part of pest management. Poorly prepared planting sites put stress on trees, and stress invariably leads to pest problems.

If you are planting on a new site that has not had Christmas trees on it before, it is usually a good idea to cultivate or treat the site with herbicides to remove vegetation that may compete with the seedlings for light, water, or nutrients. Using herbicides in the fall of the year, before planting, will also leave dead vegetation on the site that will hold the soil and reduce erosion. By controlling weeds and grasses after planting, you can continue to keep competition down, reduce mammal habitat, increase air flow, and destroy alternate hosts that some pests need to complete their life cycles.

Before planting on a site, consider potential pest problems. Are there pests on trees in nearby windbreaks or woodlots? If so, you may want to treat or remove these pest "reservoirs" before planting. If pests in the surrounding areas are difficult to manage, consider planting a tree species that can withstand injury or is resistant to the pests present.

Before planting, send soil samples to a testing facility for analysis. They can tell you if you need to adjust your pH or add nutrients to make your plantation more productive.

When replanting a harvested site, it is wise to remove or destroy residue, unsalable trees, and old stumps that might harbor or attract pests to the site.

Planting

A little extra care at planting time will pay off in good tree survival and growth. Try to avoid planting on sites that are prone to frost or drought, or where soils are stony, coarse textured, or otherwise unsuitable for trees. Such sites invite pest problems. Spacing is also important. Wider spacing will help create better air flow between trees that may reduce some needle diseases. Wider spacing and access lanes also allow more room for spray equipment, if it is needed.

Avoid planting Scotch pine next to older pines already infested with root collar weevil. (S. Katovich, USFS, Bugwood.org)

Scout fields on a regular basis. (J. O'Donnell, MSU)

When handling seedlings, keep the roots moist to ensure survival and reduce transplanting shock.

Plant your seedlings in up- and down-hill rows rather than along the contours of the land. This increases air flow and drainage, allowing trees to dry quickly after rain with less chance of pathogen spread. When you plant, spread the roots out in the planting hole to prevent them from growing in a J-shaped curve. Improperly planted, J-rooted trees are especially vulnerable to white grubs during the first and second seasons after planting. J-rooted trees also tend to be weak, unstable, and susceptible to root rot and drought damage.

Place seedlings so that the root crown is at the same level it was in the nursery bed. Pine root collar weevils injure pines more readily when the root collar is more than 2 inches below the ground because an underground "collar" is available for them to girdle.

Step 2. Monitor and Scout Trees for Pests

Even if you do a good job preparing the site and planting, insects, diseases, and other pests can still damage your seedlings and trees. Walking through your nursery or plantation on a regular basis to keep track of tree condition, pest abundance, and damage is one of the most important things you can do to maintain healthy trees. If you are observant, you can usually spot the symptoms and signs of distress before widespread damage occurs.

Scouting your trees can also help you:

- Learn which tree species and varieties are most resistant to damage
- Determine whether beneficial organisms, such as insect predators, are present
- Anticipate and prevent pest damage

- Gauge how much damage a tree can handle without a loss in grade
- Decide whether control is needed
- Judge the results of your management decisions

It is good practice to inspect your trees weekly throughout the growing season and occasionally in winter. Although most pests are active during warmer weather, some diseases are more severe in cool, moist weather, and birds and mammals do the most damage in winter when their normal food supply is scarce. When you scout, take notes and make maps of where damage has occurred. This will help you plan future scouting and control activities.

Begin pest monitoring when your stock arrives from the nursery and continue until harvest. "Hitch-hiking" nursery pests are particularly serious because even a few pests can destroy small plants and quickly spread to other parts of your plantation. To be safe, keep careful records; buy locally grown seedlings, if available; and ask your seedling supplier about guarantees and pest-free certificates.

It also pays to know something about the habits of pests you find on your trees. Certain pests, even when numerous, may not seriously damage a tree if it is large enough. For example, dozens of European pine sawflies can strip many needles off a 5-foot pine, but because they eat only old needles, the tree is barely injured and recovers fully in 2 or 3 years.

On the other hand, one or two Zimmerman pine moth larvae can kill or severely injure the same 5-foot pine in one season. Naturally, the more "significant" the pest, the more vigilant your monitoring needs to be. If your first inspection reveals no serious threats, don't stop. The situation can change quickly.

To keep abreast of local pest conditions, check into the newsletters and pest monitoring programs available for Christmas tree growers in several States. These programs can provide historical as well as current information on pest problems.

Step 3. Use a Combination of Controls

When faced with a serious pest buildup, your best strategy may be a combination of simple treatments rather than a single drastic action. Although you can use chemical control to combat pests, do not rely on it. There are many good alternatives and supplements to pesticides. The strategies suggested here are not likely to completely eliminate pests from your nursery or plantation. Instead, these strategies work to bring pest populations down to acceptable levels and keep them there. An acceptable level merely means the trees will not be degraded at the time of harvest. You can keep potential pests at acceptable levels by practicing prevention and some combination of manual, mechanical, biological, cultural, and chemical control methods.

Most of the methods presented here have been used successfully in Christmas tree

Green lacewings are important predators of aphids and scales. (W. Cranshaw, Colorado State Univ., Bugwood.org)

fields in the past. However, a few of these suggestions have not yet been extensively tested, so you may wish to try them and see if they work for you. One treatment may work well in one area and not as well in another, so continue trying new treatments or seek help if you have trouble managing a pest (see "Where to Get Help").

Manual and Mechanical Control

You can use hand methods or mechanical devices in small plantings to control pests or to make the environment unsuitable for their survival. For instance, you can sometimes hand pick low numbers of pests or knock them off the tree and crush them. You can set out fresh pieces of tree stem in the plantation to trap certain beetles. You can broadcast recorded predator calls in the field to drive off

bothersome birds. Chipping or burning trees infested with Zimmerman pine moth larvae can prevent attacks on healthy trees. Sometimes these simple controls are all that is needed to discourage costly pest damage.

Biological Control

Natural enemies, such as predators, parasites, and pathogens, can play an important role in pest control. When natural enemies become permanent residents in Christmas tree fields, pests are less likely to increase to damaging levels. The long-term nature of biological control makes it relatively inexpensive as well as environmentally safe.

Biological control can involve introducing beneficial organisms into your nursery or plantation or simply encouraging those that are already in place. These beneficial organisms include lady beetles, which devour aphids and scales by the hundreds; lacewings; spiders; and predatory mites. Parasitic wasps and flies check pest numbers by laying their eggs on the body of pest insects. And there are many diseases that can weaken or kill Christmas tree insect pests.

You can attract beneficial predators and parasites to your fields by leaving edge rows or occasional strips or clumps of certain flowering plants as a pollen and nectar source. Adult parasites of many insects need pollen or nectar for food and will search out pests in your plantation if flowering plants are available. Some of these beneficial predators and parasites can be bought commercially. You may be able to buy commercial virus preparations for sawflies such as the redheaded pine sawfly, or you can make your own. The European pine sawfly writeup in this manual includes a recipe. Once introduced, a virus persists and affects new generations of sawflies year after year.

Keep predators and parasites working for you by minimizing the use of chemical insecticides. If you must use insecticides, apply the lowest recommended dose. Apply the product at the correct time and minimize drift to reduce harmful effects on beneficial insects. Also, try to spot treat pests to further minimize

Mowed lanes discourage mammal pests, and wide spacing can have many benefits for managing pests. (R.S. Kelley, VT Dept. of Forests, Parks and Recreation)

pesticide use. You do not always need to treat an entire nursery or plantation if only a few trees or small groups of trees have been affected. In addition, there are several insecticide products that are much less toxic to predators and parasites than many of the traditional insecticides. These are sometimes referred to as biorational pesticides, and they are more likely to conserve natural enemies in your plantations.

Cultural Control

Ordinary cultural practices such as mowing, shearing, pruning, and thinning can help make your plantation less appealing to pests. If you strategically modify and time these operations, you can manipulate pest habitat to prevent and control problems even more effectively. For example, you can discourage mice by mowing the grass they

hide in. You can reduce pine needle rust by removing nearby goldenrod and aster plants. Pruning the lower branches from old trees helps control pine root collar weevil, European pine shoot moth, and some tree pathogens.

The goal is to make the habitat less favorable so pests will not multiply as rapidly. Sometimes even a slight drop in a pest's population can avert a damaging buildup. Cultural controls are among the simplest and cheapest methods available because they can complement other management operations and are environmentally safe.

Chemical Control

Chemical pesticides can be among the most effective materials used to prevent, destroy, or repel pests; because of this, they have been used too often in lieu of other control methods. If you must use chemical pesticides, it is important to choose the proper products, timing, and dosages to avoid mistakes.

Improper use of a pesticide might rid your nursery or plantation of one pest, but may very well trigger another, more serious problem. Spraying may kill natural enemies, which may lead to damaging outbreaks of mites, aphids, and scales. To maximize the benefits and minimize the hazards of pesticides, choose chemical formulations that pose the least threat to nontarget species. Adjust and calibrate application equipment so the proper amount of pesticide hits the target, and only the target. Time your treatment to avoid spray drift. We also suggest alternating classes of pesticides to reduce the chances of a pest developing resistance to a particular type of pesticide and using biorational pesticides if available.

When used as directed and in combination with other controls, pesticides can produce impressive reductions in pest populations. We have purposely left out specific product names and application rates, however, because they change so frequently. Check the label on the pesticide container for application and registration information. Extension offices and State regulatory agencies can provide up-to-date information on pesticides registered in your State.

Step 4. Evaluate Your Control Efforts

To be truly effective, pest management needs to be part of the day-to-day workings of your Christmas tree operation, from species selection to premarket inspection. This includes regular, careful scouting, even after a control treatment. By evaluating your treatments, you can decide which management techniques were successful, and which were not. You can then continue using the best techniques and minimize pest damage by design, not by chance.

Pesticides can be a viable part of an overall pest management program. (R.S. Kelley, VT Dept. of Forests, Parks and Recreation)

Using Degree Day Accumulation for Improving Timing of Insect Pest Management

Correctly timing pest management activities, such as scouting or pesticide applications, can be challenging because of variation in weather among years and among locations. For example, insect development can be 2 to 3 weeks faster in a year with warm, sunny spring weather than in a year with cold, wet spring weather.

Many Christmas tree growers and farmers know that "degree days" can be a useful tool to help time pest survey and control activities. Degree days is a term that refers to the accumulation of heat units above a threshold temperature over time. It is a convenient measure of how warm or cold it has been during the spring and summer. Monitoring degree day accumulation in your specific area can help you estimate when specific insect pests are likely to be present. See table 1 for degree day information for a number of Christmas tree insect pests.

Table 1. Growing Degree Days for Selected Christmas Tree Insects[1] – *This information is not available for all insect pests.*

Insect	Life stage	GDD_{50}[2]
Balsam gall midge	egg hatch	60-100
	adults laying eggs	150-300
	galls apparent	550-700
Balsam twig aphid	stem mothers present (control target)	100-150
Black pineleaf scale	egg hatch	1068
Cooley spruce gall adelgid	1st adults active - Spruce (control target)	25-120
	1st galls visible - Spruce	200-310
	1st adults active - Douglas fir	90-180
	1st nymphs - Douglas fir (control target)	90-150
	2nd nymphs - Douglas fir (control target)	600-1000
	2nd adults active (control target)	1500-1600
Douglas-fir needle midge	Emergence of adults	200-225
Eastern pine weevil	1st adults active	25-100
	2nd adults active	1200-1400
Eastern pine shoot borer	1st adults active	75-200
Eastern spruce gall adelgid	1st adults active (control target)	25-100
	egg hatch, galls begin forming	250-310
	2nd adults active (control target)	1500-1600
European pine sawfly	1st larvae	100-195
European pine shoot moth	1st larvae	50-220
	adults active	700-800
	egg hatch	900-1000
Gypsy moth	egg hatch, 1st larvae	145-200
	young caterpillars	450
	pupation	900-1200

Insect	Life stage	GDD_{50}[2]
Introduced pine sawfly	1st larvae	400-600
Jack pine budworm	young larvae feeding	300-350
	large larvae feeding - defoliation apparent	650-700
Jack pine sawfly	eggs; young larvae	100-200
	larger larvae consuming needles	275-500
Japanese beetle	adults emerge and feed	950-2150
Juniper scale	egg hatch	550-700
Pales weevil	1st adults active	25-100
	2nd adults active	1200-1400
Pine chafer (Anomala beetle)	1st adults active	450-600
Pine engraver (Ips bark beetle)	1st adults active	100-150
Pine needle midge	1st adults active	400-500
Pine needle scale	1st generation egg hatch	250-400
	1st generation - hyaline stage (control target)	400-500
	2nd generation egg hatch	1250-1350
	2nd generation - hyaline stage (control target)	1500
Pine root collar weevil	1st adults active	300-350
	2nd adults active	1200-1400
Pine shoot beetle	optimal control window	450-500
	new adults emerge; begin shoot-feeding	500-550
Pine tortoise scale	egg hatch begins; 1st crawlers	400-500
	egg hatch ends; last of the crawlers	1000-1200
Pine tube moth	adults; egg laying; caterpillars	90-250
Redheaded pine sawfly	1st larvae	400-600
Spruce bud scale	egg hatch, 1st crawlers	700-1150
Spruce budworm	1st larvae	200-300
Spruce needleminer	1st larvae	150-200
Spruce spider mite	1st egg hatch	150-175
Striped pine scale	egg hatch	750-800
Turpentine beetle (bark beetle)	parent beetles colonizing brood material	300-350
White pine weevil	1st adults active	25-220
	2nd adults active	1200-1400
Zimmerman pine moth	1st larvae	25-100
	adult flight	1700

[1] Michigan State University Extension *http://www.ipm.msu.edu/agriculture/christmas_trees/gdd_of_conifer_insects*

[2] GDD_{50} = Growing Degree Days using a 50° F threshold (basis)

In the North Central and Northeastern regions of the United States, we typically use 50 °F as a threshold (also referred to as a base temperature) for degree day accumulation. Because insects are cold-blooded animals, temperatures usually must be relatively warm before feeding, flight, egg hatch, or other important activities can occur. Development of immature insects is especially affected by temperature.

In spring, as the weather warms up, temperatures begin to exceed 50 °F for at least a few hours during the day. As spring progresses and it continues to get warmer, temperatures exceed 50 °F for an increasing portion of the day. In its simplest form, one degree day is equivalent to 24 hours of temperatures at 51 °F, two degree days are equivalent to 24 hours of temperatures at 52 °F, and so on. A warm day generates many more degree days than a cool day, even though both might be over 50 °F. By the end of summer, more than 1,200 degree days may accumulate in northern areas of the North Central region, while 2,000 degree days or more may accumulate in the southern portion of this region.

Agricultural meteorologists from most land grant universities monitor weather and use more sophisticated methods to calculate current degree day accumulations throughout the growing season in various locations around their State. Meteorologists can also use weather forecasts to project future degree day accumulations. Degree day information is typically available on university Extension Web sites or in weekly publications. You can also access other Web sites, such as The Weather Channel, for degree day information for a specific region.

Making sound pest management decisions can increase the likelihood of producing high-quality Christmas trees. (J. O'Donnell, MSU)

Where to Get Help

Because so many pests cause look-alike symptoms, it is sometimes difficult to pinpoint the pest causing damage in your plantation. When in doubt, locate the appropriate diagnostic laboratory in your State to help identify the pest. Most of these facilities are plant disease or pest diagnostic labs located at land grant universities. There is generally a fee associated with these services.

Extension offices can be an excellent resource for management recommendations, including information on pesticides registered in your area. Christmas tree consultants may also be available to provide many of these services.

The Internet can also be a source of helpful information. Many land grant universities and State agencies maintain Web sites with photos and information to help you identify pests affecting trees. Be careful to check the source of information; not everything on the Internet is reliable.

Some of the best advice may come from other growers who have dealt with a pest problem similar to yours. Joining a Christmas tree growers' association is a good way to connect and compare notes with growers in your area.

How to Submit Materials for Identification

Submitting good-quality samples and providing information on local conditions can be a great help in the identification process. Each diagnostic laboratory will have a general form requesting some key pieces of information and a list of items to follow when collecting, packing, and submitting sample material. Follow those directions. It is always a good practice to contact a lab prior to submitting material to make sure the lab will be staffed on the day your sample arrives.

When collecting material to submit, make sure you also note these things:

- Location of the plantation
- Site conditions (wet, dry, etc.)
- Age and species of affected trees
- Part(s) of tree damaged
- Pattern of damaged trees in the plantation (scattered or grouped)
- Date of symptom development
- Extent of damage (number of trees or acres affected)
- Management history (fertilization, pesticide and herbicide use, etc.)

Digital Images

Having quality, high-resolution photographs can sometimes provide the information needed to make a reliable identification. It is helpful to take photographs from a variety of perspectives, from close-ups of damaged tree sections to images that show the entire tree and surrounding landscape.

Needle Discoloration or Distortion

Scattered, single needles or clumps of needles may be spotted, banded, stippled, or totally discolored—yellow, red, brown, or black. You may find fruitbodies, swellings, or scales on injured needles. Some may be distorted. If needles are chewed off, see next section. If shoots, branches, or entire trees are discolored, see other injury categories.

Admes Mite

Eurytetranychus admes

Hosts: Black Hills, white, and Norway spruce; occasionally Colorado blue spruce

Importance: Severe attacks degrade Christmas trees. However, this mite is generally not as commonly encountered nor as damaging as the spruce spider mite.

Look For:

- *Yellow stippling or bronzing of the foliage*, similar to damage caused by other spider mites. When damage is heavy, some of the needles turn brown and eventually drop.

- Adults can be found resting on the underside of needles with their legs stretched out in front and back.

- Admes mites have a *dark reddish-brown body with light-tan legs*; they are larger than the spruce spider mite.

- When immature mites molt, they line up in a row on a needle. *The row of exoskeletons left behind is visible and looks like white flocking on the needles.*

- *Eggs are red and can be seen with the naked eye.* Under a microscope you can see shallow but distinct ridges running from the base to the center of the egg.

Pests that cause similar symptoms: Spruce spider mites, eriophyid mites, Rhizosphaera needlecast

Biology: This mite has several generations a year. Mites overwinter as eggs on the host. Most eggs will be found at the base of needles. Adults can be found on the underside of the needles. Mites feed on tree sap and cause small feeding wounds on needles. Several natural enemies feed on spider mites including small, black lady beetles; lacewing larvae; minute pirate bugs; and predatory mites.

Monitoring and Control: Inspect trees of all ages throughout the growing season. Because mite predators are extremely susceptible to many insecticides (especially conventional insecticides), do not apply an insecticide until the health or appearance of the plant is threatened. If treatments are required, spray only the trees that are severely infested and use a registered miticide. This will help conserve the beneficial natural enemies of the mites.

Next Crop:

- If this is a serious problem in your area, consider growing something other than spruce.

Adult mite. (R. Lehman, PA Dept. of Agriculture, Bugwood.org)

Reddish colored egg. (J. O'Donnell, MSU)

Bronzing of needles. (J. O'Donnell, MSU)

Air Pollution Injury

Air Pollution Injury

Species Affected: All conifers, especially eastern white pine

Importance: Air pollution damage reduces growth, causes early needle loss, increases vulnerability to diseases and insect pests, and occasionally kills trees.

Look For:

- *Yellowing, stunting, and early shedding of old needles.*
- *Yellow, red, or brown tips on current-year needles.*
- *Yellow flecks, stipples, or bands on needles.*
- *Injury on nearby broad-leaved plants*, for example, dead tissue at leaf margins (fluorides), or between leaf veins (sulfur dioxide), or stippling (ozone). Aspen, birch, alfalfa, and sweet corn are particularly sensitive to air pollution injury.

Factors that cause similar symptoms: Drought, herbicide injury, nutrient deficiencies

Biology: Air pollutants produced by automobiles (ozone), industrial processes (fluorides), and coal- and oil-burning factories (sulfur dioxide) will injure a wide range of plants in or near Christmas tree plantations. The amount of damage to conifers depends on the age of the needles, genetic makeup of the tree, pollutant concentration, weather, and how long the tree was exposed. Needles are most susceptible when elongating during early summer.

Monitoring and Control: Carefully examine affected trees to rule out injury by other pests. Remove dead shoots and trees to prevent a buildup of other pests on this material.

Next Crop:

- Before choosing a new plantation site, check the surrounding area for industries that may produce damaging pollutants. Most damage occurs within 10 miles of these sources; however, ozone injury can also occur in remote areas.
- Plant seedlings that are genetically resistant to air pollution injury. For example, spruces are resistant to sulfur dioxide, ozone, and fluorides; balsam fir, Fraser fir, Douglas-fir, and red pine are resistant to ozone.

Yellow and brown tips on current-year needles. (MN-DNR Archive, Bugwood.org)

Banded, stunted needles. (P. Kapitola, State Phytosanitary Administration, Bugwood.org)

Balsam Gall Midge

Paradiplosis tumifex

Hosts: Balsam fir and Fraser fir

Importance: Feeding by the larvae of this tiny fly causes small galls to form on new needles. Galled needles drop prematurely in late summer or fall, which leaves bare spots on branches that lower the market value of affected trees. Shearing can remove some affected shoots, and injured trees will recover if they are not heavily infested again for 3 or 4 years.

Look For:

OCTOBER TO APRIL

* *Thin foliage and bare branches*, particularly near the upper crown.

MAY

* *Adult female midges laying clusters of eggs* under bud scales or between the needles of newly emerged shoots. Adults are orange and resemble mosquitoes in size and appearance.

JUNE TO OCTOBER

* *Galls: globe-like swellings near the base of new needles.* In heavy infestations, there may be several galls per needle; more than 90 percent of new shoots may bear infested, galled needles.

* *Small, yellow-orange or pink larvae inside galls*. There will be only one balsam gall midge larva in a gall; if two larvae are present, then one is a beneficial "inquiline" midge species (see explanation below in **Biology**).

Biology: Mature larvae drop from the needles in autumn, overwinter in litter or soil beneath the tree, then pupate in spring. Winged adults emerge in late May to early June. Females lay eggs on the bud scales and newly emerging foliage. Larvae feed on the current-year needles, causing galls to form.

An inquiline midge (*Dasineura balsamicola*) is an important cause of balsam gall midge mortality. An "inquiline" refers to an insect that lives with another species. This midge has a life cycle similar to balsam gall midge but does not cause galls to form. If at least one inquiline midge larva is present in a gall, the balsam gall midge larva will die.

Larva. (D. Carlton, CFS, New Brunswick)

Monitoring: See table 1 (page 22) for degree day information.

ADULTS: Place emergence traps beneath previously affected trees to determine when adult midges are laying eggs. Traps can consist of either a bottomless, small wooden box or an opaque plastic flower pot. The trap should have a small hole in the side with a clear vial exposed to the light. Adults will be attracted to the light and can easily be seen in the vial. Set traps out in late April or early May beneath the drip line of a tree. Bury the trap up to 1 inch deep. For best results, place a trap beneath 10 trees in a field where damage occurred in previous years.

GALLS: Look for galls anytime between June and October starting 3 to 4 years before harvest. Infestations tend to be heaviest in the upper crowns of individual trees.

Globe-like galls on needles. (R.S. Kelley, VT Dept. of Forests, Parks and Recreation, Bugwood.org)

Egg-laying adults. (D. Carlton, CFS, New Brunswick)

Control:

- Remove and burn heavily infested branches or trees in mid to late summer before larvae and needles drop to the soil. Removing affected trees in early spring will only encourage emerging adults to lay eggs on nearby and perhaps previously uninfested trees.

- Kill adult midges by spraying heavily infested trees with a registered insecticide within 7 days of adult emergence. Avoid spraying trees in areas where only a few midges are emerging or where only a few galls were present. This will conserve the natural enemies that keep midge populations in check.

- Once galls are formed, insecticides will not be effective.

Next Crop:

- Avoid planting balsam fir or Fraser fir where balsam gall midge has consistently been a problem.

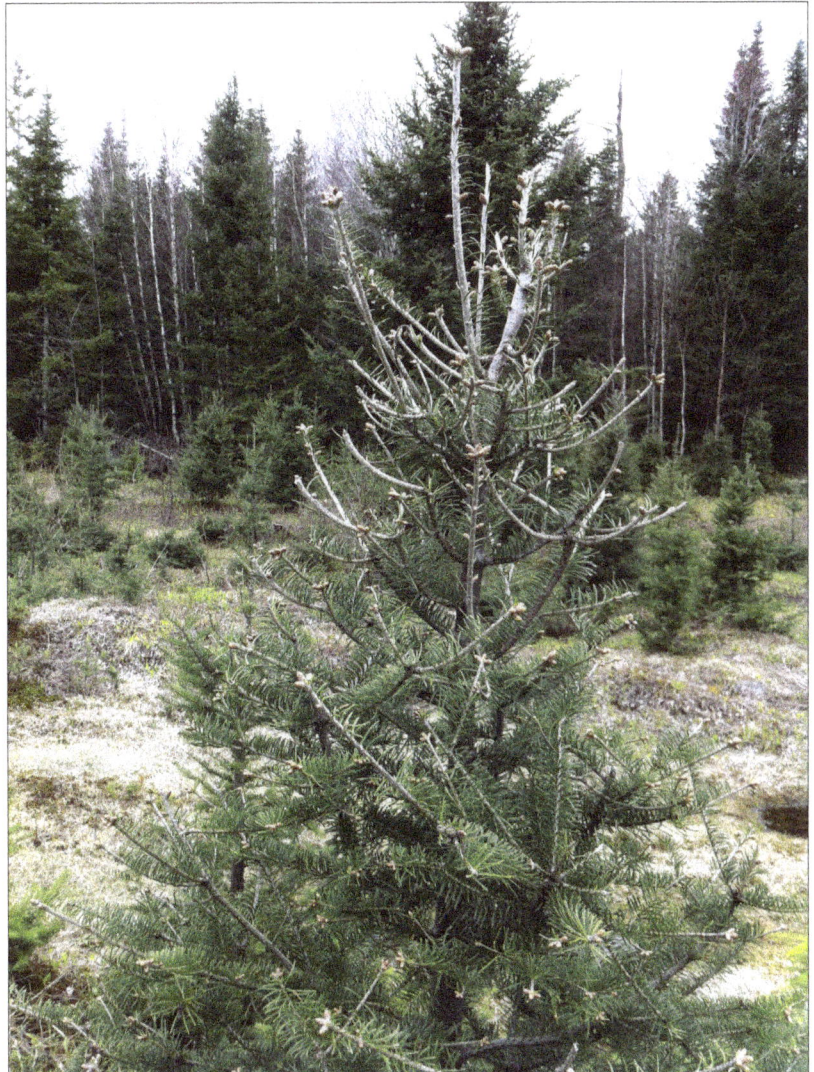
Thin foliage and bare branches. (D. Carlton, CFS, New Brunswick)

Emergence trap. (D. Carlton, CFS, New Brunswick)

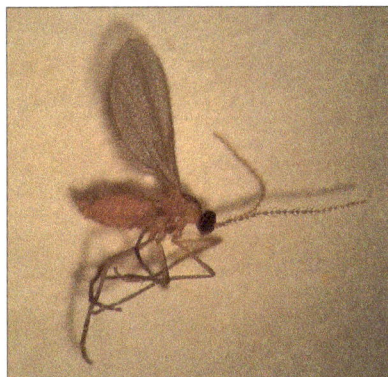
Adult male. (D. Carlton, CFS, New Brunswick)

Eggs. (D. Carlton, CFS, New Brunswick)

Balsam Twig Aphid

Mindarus abietinus

Hosts: Balsam fir, Fraser fir, other true firs

Importance: Balsam twig aphids feed on sap of swelling buds and developing needles early in the season. This feeding reduces needle growth and causes needles to curl and become distorted. Needles outgrow most of the curling during the summer. Severe infestations can reduce tree appearance.

Look For:

JULY TO MARCH

- *Curled and stunted needles from the previous year.* Shoots may appear twisted and silvery.

APRIL TO EARLY MAY

- *Newly hatched nymphs on or near the maturing buds.* Newly hatched nymphs are white or yellow and may be on the underside of the previous year's needles, usually within 2 inches of buds.

MAY

- As buds swell and begin to break, *feeding causes the needles to curl and appear to wrap around the shoot.*

- *White, "woolly" wax and honeydew* may be seen on needles.

MAY TO JUNE

- *Green-winged and wingless aphids covered with powdery wax* may be seen on the expanding, current-year shoots.

- *Bees, wasps, and ants* may feed on the honeydew. Adult and larval lady beetles, lacewings, and larvae of syrphid flies (also called hover flies or flower flies) may be seen feeding on aphids.

Biology: Overwintering eggs hatch in spring and nymphs quickly mature into a stage called "stem mothers." Each stem mother produces 20-40

Curled, current-year needles. (S. Katovich, USFS, Bugwood.org)

offspring. These aphids feed on sap in the tender, current-year needles and are the only stage that causes appreciable damage. They eventually mature into winged adults and may disperse to other trees. A final generation in mid or late summer will produce the overwintering eggs, but do little feeding.

Monitoring and Control:

Evaluate the extent of damage in autumn or in early spring before bud break. Check upper, mid, and lower canopy shoots. Determine whether damage warrants application of an insecticide. Studies have shown that damage from balsam twig aphid did not affect Christmas tree customers at choose-and-cut farms or wholesale buyers until at least 30 percent of the shoots had been damaged.

If an insecticide application is needed, target the stem mothers for control. The ideal time to spray is after the stem mothers have hatched but before their offspring begin to feed. In spring, scout previously damaged trees to determine if the stem mothers have hatched. Use a wooden dowel and rap mid-canopy shoots over a small, black cloth. Black cloth held in an embroidery ring, for example, works well. If the white to greenish stem mothers are present, they will be readily visible. Sample aphids on both the south- and north-facing sides of trees. Monitoring degree day accumulation can also help with effective spray timing.

- See table 1 (page 22) for degree day information.

- Spraying trees after the current-year needles have begun to curl has little effect on damage or aphid numbers.

- Avoid spraying trees or surrounding vegetation with insecticides when helpful aphid predators are present.

- Balsam twig aphid may be more likely to infest species or individual trees that break bud early. Many balsam fir varieties break bud before Fraser fir trees. Prioritize early bud-breaking trees for scouting and monitoring. Trees that generally break bud later can often be retained until they reach a larger size.

- Nitrogen fertilizer applied in spring tends to increase aphid numbers.

Next Crop:

- Conserve beneficial predatory insects that will prey on aphids by providing vegetation, including flowering plants, along field borders.

Nymphs. (E.B. Walker, VT Dept. of Forests, Parks and Recreation, Bugwood.org)

Adult balsam twig aphid. (R. Lehman, PA Dept. of Agriculture, Bugwood.org)

Brown Spot Needle Blight

Mycosphaerella dearnessii

Host: All pines

Importance: The browning and early needle loss caused by this fungus results in affected pines being unsalable as Christmas trees.

Look For:

- *Reddish-brown, resin-soaked spots with yellow margins* on the needles in late summer. Spotted needles turn yellow, then brown.

- *Black fruitbodies* flush with the surface of dead, dry needles on the tree or on the ground. These fungal structures protrude from the needle when wet.

AUGUST TO OCTOBER

- *Brown needles,* especially on the lower branches and on the moist, shaded north side of trees. Needles turn brown from their tips towards their bases.

MAY TO JULY

- *New shoot and needle growth* on the tips of branches that hold dead, brown needles. Most dead needles fall off, leaving only tufts of new, green growth on the branch tips.

Pests that cause similar symptoms: Lophodermium needlecast, pine needle scale, winter injury

Biology: Prolonged wet periods, particularly during June and July, provide favorable conditions for infection. Old needles are more resistant to infection than young ones.

Monitoring and Control:
Inspect trees of all ages at least once between August and October. Look for needle spots and browning on current-year and older foliage on the lower branches. If trees exhibit disease symptoms, consider treating the entire plantation the following spring. If the disease occurs in small pockets, treat only the affected and surrounding trees.

- Cut and immediately remove small pockets of severely diseased trees. Treat trees within 30 feet of diseased trees with a registered preventive fungicide.

- Promote good air movement by controlling weeds and pruning lower branches.

- Do not shear during wet weather because spores could be carried to healthy trees on shearing tools.

- Shear healthy trees first; disinfect tools after shearing.

- Do not leave live branches on stumps of harvested trees because the stumps can serve as disease reservoirs.

- Apply a registered, preventive fungicide to trees when needles are about half grown. Repeat two or three times, once every 2 to 3 weeks, to protect new and old foliage.

Next Crop:

- Plant only disease-free stock. If you suspect infection, have seedlings examined by a pest specialist.

- Plant disease-resistant varieties of Scotch pine, such as the long-needled varieties from Central Europe.

- Plant more than one pine species or variety so that one disease will not damage the entire crop.

- Do not plant Scotch pine seedlings next to Scotch pine windbreaks that could be disease reservoirs.

Reddish-brown spots with yellow margins on infected needles. (D.D. Skilling, USFS, Bugwood.org)

Black fruitbodies on infected, dead needles. (D.D. Skilling, USFS, Bugwood.org)

Brown, infected needles on lower branches. (USFS - NCRS Archive)

Cyclaneusma Needlecast

Cyclaneusma Needlecast (= Naemacyclus Needlecast)

Cyclaneusma minus

Host: Scotch pine

Importance: The early yellowing and loss of older needles caused by this fungus weaken and degrade trees.

Look For:

SEPTEMBER

- *Light-green spots on 2- and 3-year-old needles.* Spots enlarge and lighten in color, and needles eventually turn yellow, then brown.
- *Yellow needles with dark-brown, horizontal bands.*

OCTOBER TO MAY

- *Shedding of yellow needles* anywhere on the tree.

- *Off-white, waxy fruitbodies on brown needles* that are most noticeable in wet weather when they protrude from the needle due to swelling.
- Pests that cause similar symptoms: *Fall needle drop, pine needle scale, winter injury*

Biology: Needles of all ages are susceptible to infection. Most trees are infected between mid-April and late June, but infection is possible through December. *Cyclaneusma* spores spread most readily after rainfall. Development of visible symptoms on needles may depend on host vigor.

Monitoring and Control: Inspect trees of all ages in late fall and early spring. Consider treating the entire plantation in early spring if disease incidence is high.

- Apply a registered, preventive fungicide three times, once every 2 to 3 weeks between mid-April and late June, starting before Scotch pine buds open. Continue spray schedule into late fall for complete control.
- Control weeds and maintain tree spacing that allows good air movement between trees.

Next Crop:

- Buy planting stock from a nursery that uses preventive treatments for all diseases.
- If available, plant tree stock from sources with genetic resistance to *Cyclaneusma.*
- Avoid planting next to old Scotch pine windbreaks or plantations that can be disease reservoirs.
- Choose a planting site that promotes drying of wet foliage.

White, waxy fruitbodies on dead infected needles. (J. O'Brien, USFS, Bugwood.org)

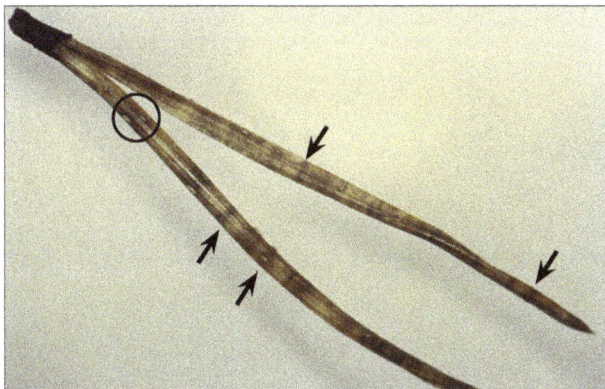

Infected needles with dark bands (arrows) and fruitbody (circle). (J. O'Brien, USFS, Bugwood.org)

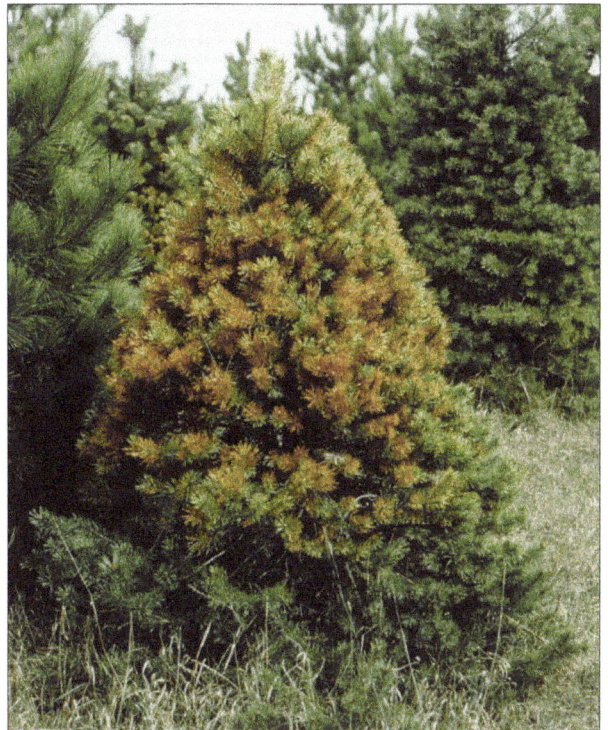

Yellow-brown needles on a diseased tree. (USFS - FHP - St. Paul Archive, Bugwood.org)

Dothistroma Needle Blight

Mycosphaerella pini

Host: Austrian and red pine

Importance: This fungus infects and kills needles of all ages. Severely affected trees can be killed or may become more susceptible to other diseases.

Look For:

FALL

- *Yellow to tan needle spots* that enlarge to form distinct brown to reddish-brown bands.
- *Dead needle tips* beyond the reddish-brown bands. Needle base remains green.
- *Black fruitbodies* in dead spots or bands on needles.

Pests that cause similar symptoms: Pine needle scale, winter injury

Biology: Spores spread by wind and rain can infect needles throughout the growing season. However, new needles are not susceptible until they have emerged from the needle sheaths in early summer. Fruitbodies appear in the fall, and spores are released the following spring and summer.

Monitoring and Control: Check trees of all ages in the fall. If you find needle spots on any of your trees, consider treating the entire plantation the following year. Take other preventive measures immediately to avoid spreading the pathogen.

- Do not shear during wet weather because spores could be carried to healthy trees on shearing tools.
- Shear healthy trees first; disinfect tools after shearing.
- Do not ship infected seedlings or Christmas trees because this is how this fungus is spread to new areas.
- Apply a registered fungicide once between mid-June and mid-July to protect all foliage. For complete control, consider spraying once in mid- to late May and again in mid-June to mid-July.

Next Crop:

- Plant only disease-free nursery stock.
- Avoid planting Austrian pine. Plant disease-resistant Austrian pine varieties. Trees from a Yugoslavian seed source have exhibited resistance to Dothistroma needle blight.
- Do not plant Austrian pine near windbreaks of Austrian pine that can serve as reservoirs of the disease.

Dead needle tips. (USFS - NCRS Archive)

Black fungal fruitbodies. (P. Kapitola, State Phytosanitary Administration, Bugwood.org)

Douglas-Fir Needle Midge

Contarinia pseudotsuga

Hosts: Douglas-fir

Importance: Douglas-fir needle midge is a small fly. Larval feeding inside current-year needles causes the needles to discolor, swell, and bend. Infested needles drop from the tree in the fall, usually just before harvest. Severe infestations can cause unacceptable needle loss.

Look For:

MAY TO JUNE

- *Adult midges*, about the size of a mosquito.
- *Orange eggs* laid between expanding bud scales in early spring. A hand lens may be required to see the eggs.

JULY TO OCTOBER

- *Discolored patches on infested needles* that initially appear yellow, then purple, and finally brown.
- *Swollen areas on needles* may resemble a gall. Larvae feed inside these swollen areas.

- *Needles may bend or appear distorted.*
- *Needle drop, often late in the fall.*

Pests that cause similar symptoms: Rhabdocline needlecast symptoms can resemble needle midge damage. Feeding by Cooley adelgids can discolor and distort Douglas-fir needles.

Biology: Larvae, the immature stage of the midge, feed in the needles throughout the summer. A single needle may harbor several larvae. The larvae complete their feeding in autumn and drop to the soil beneath infested trees, where they will spend the winter. In early spring, the larvae pupate. Adults emerge from the soil in mid- to late May and mate, and females lay eggs in the soft needles of the expanding buds. Eggs hatch within a few days, and the midge larvae immediately bore into the growing needles to feed.

Monitoring and Control: Insecticides can be used to

control Douglas-fir needle midge, but the timing of the application is critical. Use yellow sticky traps or box traps to determine when adult midges are emerging in the spring. Set out at least 10 traps per field by mid-April. Check the traps daily once your area approaches the degree day accumulation required for emergence to begin.

- See table 1 (page 22) for degree day information.
- Apply a registered insecticide when adults begin to emerge in spring. Adult midges must encounter treated foliage before they lay eggs for control to be effective.
- Late application of insecticide results in little or no control and can worsen midge problems by killing parasites that emerge later.

Next Crop:

- Avoid planting Douglas-fir where this midge has become a perennial problem.

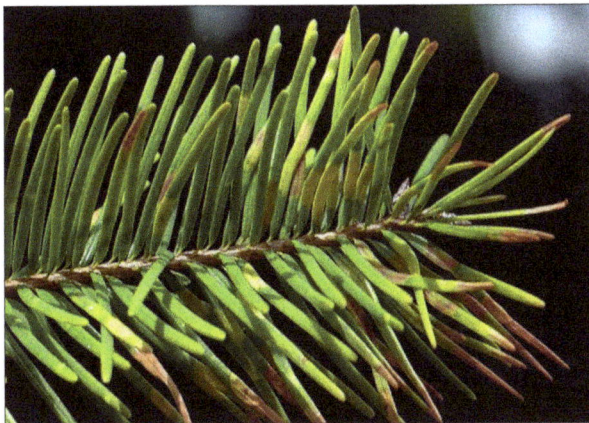

Discolored patches on infested needles. (W. Strong, BC Ministry of Forests, Bugwood.org)

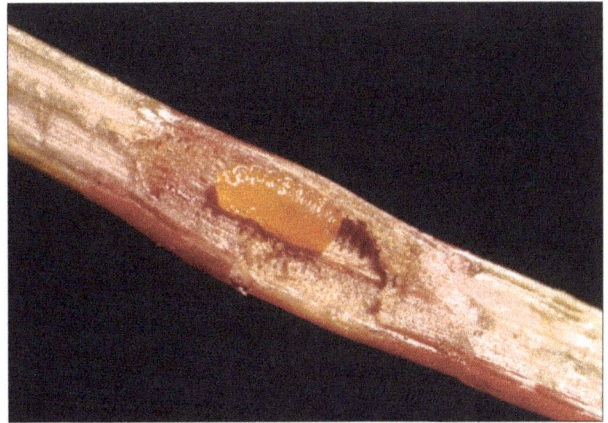

Larva found inside swollen area on a needle. (S. Tunnock, USFS, Bugwood.org)

Drought Injury

Species Affected: All trees

Importance: Drought-stressed trees lose foliage, grow slowly, and become more susceptible to insect pests and diseases. Severe drought may kill trees of all sizes, but newly planted seedlings are most susceptible.

Look For:

- *Wilting, dying needle tips and discolored foliage* on the top branches. Symptoms may not appear until a year or more after trees have been stressed by drought.

- *Wilting of the current year's shoots.*

- *Dead tree top, short needles, and sparse foliage.* These indicate a general decline in vigor that becomes evident in the years following drought.

Pests that cause similar symptoms: Scattered dead seedlings and trees can be caused by Armillaria root rot, Diplodia canker, white grubs, and pocket gophers. Needle and branch injury can be caused by eriophyid mites, herbicide injury, air pollution, Leucostoma canker, and Rhizosphaera needlecast. Drought stress often encourages many of these pests, especially Armillaria root rot and Diplodia canker.

Biology: Drought stress occurs when trees need more moisture than is available in the soil. This condition may be caused by one growing season of severe drought or several seasons of below-normal rainfall. Young trees are especially sensitive because their root systems are less extensive than those of older trees. In addition, young trees with existing root problems often are the first to die during periods of drought, including trees that were poorly planted and developed J-roots or other root problems.

Monitoring and Control: Inspect stressed trees for attack by invading pests during and after periods of drought. Follow control recommendations for each pest as needed.

- Control weeds and grasses in and around plantations to reduce competition for water during dry periods.

- If drought conditions persist, irrigate to replace soil moisture.

- Remove and chip or burn all dead trees as soon as possible; they may encourage other pest problems.

Next Crop:

- Avoid planting on very sandy, drought-prone sites.

- Do not plant shallow-rooted species, such as true firs and spruces, on drought-prone sites.

- Remove weeds before planting.

- Monitor planting closely; poorly planted trees are very susceptible to drought injury.

Wilted, dying current-year shoots. (USFS - NCRS Archive)

Dead tree top. (USFS - NCRS Archive, Bugwood.org)

Elongate Hemlock Scale

Fiorinia externa

Hosts: Hemlock is the favored host; firs and spruces are readily infested; less common on pines, Douglas-fir, and yews

Importance: An introduced pest. This scale is currently found over much of the Eastern United States. When infestations are heavy, trees appear yellow and growth can be stunted. Premature needle drop, branch death, and tree death can occur. Male insects excrete a white, waxy coating on the top surface of needles that can make trees unsalable.

Look For:

- *Yellowing of needles*, especially on interior needles on lower branches.
- *Scale coverings on the undersurface of needles*; scale coverings are longer than they are wide.
- *Brown female scales and white male scales*; both can be found on a needle.

SUMMER

- *Infested trees may appear faded in color or dirty from a distance.* Adult males excrete a white, waxy coating on needle surfaces.

SPRING TO FALL

- *Yellow crawlers and immature scales.* Crawlers have six visible legs; immature scales are not elongated.
- *Tiny adult males with wings, prominent black eyes, and long antennae.*

Pests that cause similar symptoms: Spruce spider mite

Biology: This pest has one generation per year in the Northern United States; two generations can occur further south. Overwintering can occur in different life stages, and egg hatch and crawlers can be found throughout the growing season. This staggered life cycle can complicate the timing of treatments. Crawlers settle on the underside of needles and feed by inserting straw-like mouthparts. They remain in place and form a translucent, waxy covering. Female scales mature over a 6- to 8-week period and remain alive for up to a year. Tiny winged males fly to the immobile female scales and mate. Females develop 12-16 eggs under the waxy scale.

Monitoring and Control: Summer scouting is useful since males create a white, cottony material on needle surfaces that makes trees look flocked. Confirm the presence of elongate hemlock scale by looking for scales on the undersurface of needles. If the scale is detected early, treatments can be limited to individual infested trees. Apply insecticides with care since you do not want to eliminate natural enemies of this insect.

- Apply registered insecticides with a high-pressure sprayer that creates a fog or mist that moves material onto the undersurface of needles.
- It can be difficult to determine if scales are dead following treatments; therefore, maintain an active monitoring program.
- Remove and destroy heavily infested trees.

Next Crop:

- If this is a serious problem in your area, consider growing other species.
- Avoid planting susceptible trees next to infested hemlock.
- Use wide spacing; this may limit scale movement between trees.

Scales on the undersurface of hemlock needles. (E.R. Day, VA Polytechnic Institute and State Univ., Bugwood.org)

Brown female and smaller white male scales. (K. Abell, Univ. of MA, Bugwood.org)

Eriophyid Mites
(also called rust or sheath mites)

Nalepella spp., *Setoptus* spp.

Hosts: All pines, firs, spruces, hemlock

Importance: Immature and adult eriophyid mites feed on the needles and buds of many conifers, causing foliage to become discolored and distorted. Severe attacks degrade Christmas trees.

Look For:

- *Blotchy, pale-yellow, stippled needles.*

- Shoots of infested trees may have an *unusually large number of buds.*

- *Needles may be twisted or hooked.*

APRIL TO OCTOBER

- *Tiny, cream-colored mites with a carrot-shaped body* inside needle sheaths (sheath mites) or on the undersurface of needles (rust mites). For sheath mites, pull the needle cluster apart until the papery needle sheath splits, then closely examine the lower part of the needle with a hand lens.

Factors that cause similar symptoms: Spruce spider mite, drought, herbicide injury

Biology: (Sheath mites) Several overlapping generations of eggs are laid on the needle sheaths, starting when the weather warms. Mites feed on the tree's sap under the needle sheaths.

High numbers of mites result in discolored needles.

(Rust mites) Eggs overwinter on host trees and hatch very early in the spring. Rust mites are cool-season mites; they are most active in early spring and late fall. There are multiple generations each year.

Mite populations can swell when their natural enemies are inadvertently killed by repeated use of insecticides to control other pests.

Monitoring and Control: Inspect trees of all ages throughout the growing season. Damage can show up late into the fall. The best time to treat for eriophyid mites is mid-May to mid-June before needles elongate fully.

- Drench infested trees or spray with a registered pesticide that is labeled to kill eriophyid or rust mites anytime between May and October to kill adults.

Depending on what product is used, a second application 10 to 14 days later may be needed to kill newly hatched mites.

- Limit the use of insecticides to avoid killing mite predators.

Blotchy, pale-yellow needles. (USFS - NCRS Archive)

Mite infestation on needles (A.J. Boone, SC Forestry Commission, Bugwood.org)

Fall Needle Drop

Species Affected: All Christmas tree species

Importance: Fall yellowing of inner foliage is a natural occurrence and does not harm Christmas trees before or at harvest time. It is especially noticeable on eastern white pine and some varieties of Scotch pine.

Look For:

SEPTEMBER TO NOVEMBER

• *Yellowing and browning of the oldest foliage throughout the crown of the tree.*

Pests that cause similar symptoms: air pollution, various needlecast diseases

Biology: All conifers shed their oldest needles each year. These needles turn bright yellow or brown in September or October and drop off at or before the harvest period. Each species tends to keep its needles for a defined length of time. White pines generally have 3 years of needles in the summer, drop the oldest needles in the fall, and hold 2 years of needles in the winter. Austrian and Scotch pines generally hold 3 years of needles during the winter. A healthy fir or spruce should retain 4 to 6 years of needles. Because of these differences, fall needle drop is more obvious on pines than on spruces and firs.

Monitoring and Control: Not necessary.

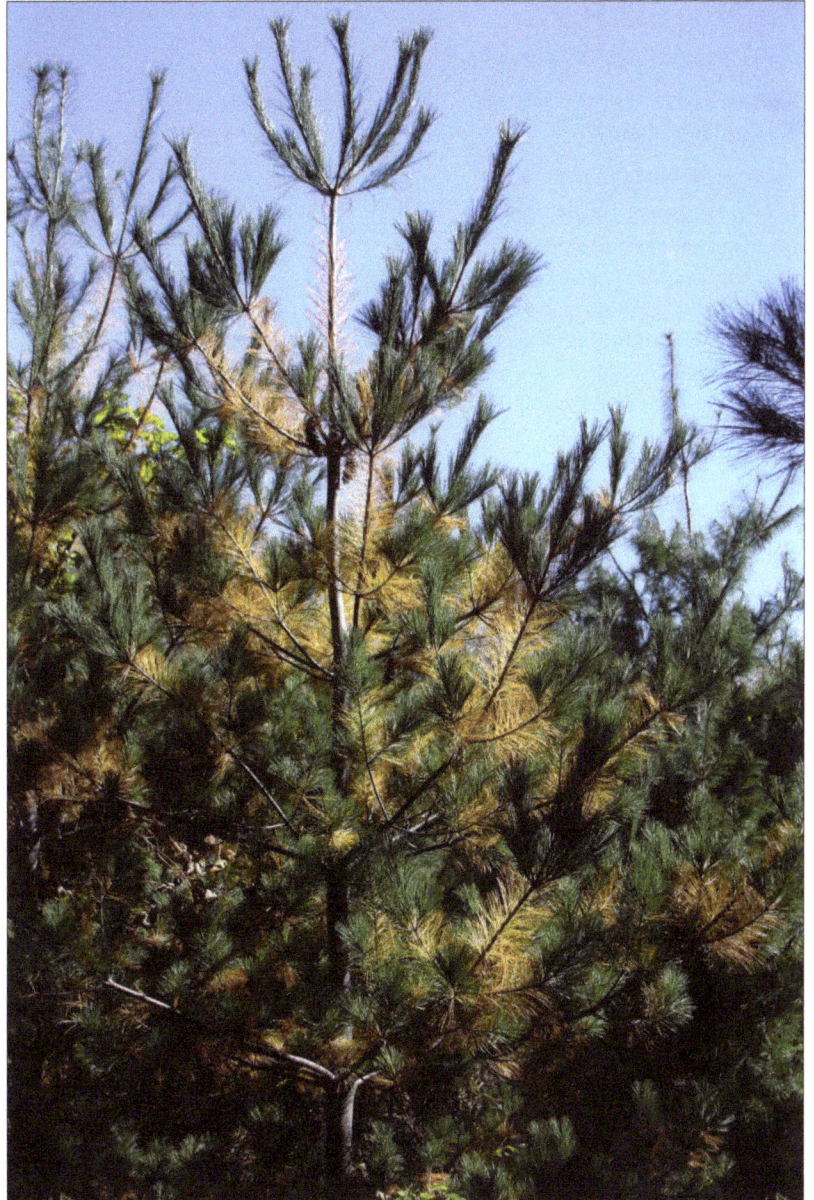

Fall yellowing of old needles. (S. Katovich, USFS, Bugwood.org)

Fir Needle Rust

Uredinopsis spp., *Milesina* spp.

Hosts: Balsam and white fir, potentially Fraser fir

Alternate Host: Ferns

Importance: Current-year needles that are infected turn reddish brown, shrivel, and drop to the ground before the end of the summer. Infected trees will have sparse foliage. Several years of repeated infection can reduce tree quality.

Look For:

JULY TO AUGUST

- *Yellow current-year needles with white pustules on the underside.*

Biology: The balsam fir needle rust fungus needs two hosts, fir and ferns, to complete its life cycle. Current-year needles of fir become infected as shoots are elongating. As the growing season progresses, infected needles turn yellow and produce spores in white pustules. These spores infect various species of fern (depending on the species of rust). The fungus overwinters on dead fern leaves and produces spores in the spring that are disseminated by wind back to the fir. Infection of fir trees requires cool, moist conditions during shoot elongation. If weather conditions are too dry, infection will not occur.

Monitoring and Control: Examine your trees from July through mid-August. Consider managing the alternate host if you observe high disease incidence. This disease usually causes only a slight loss of the current year's foliage. Infection is closely tied to specific weather conditions and will not occur every year.

- If disease incidence is high, mow or kill ferns in the plantation. Since spores are produced in the spring on dead fern leaves, expect lower levels of disease one year after ferns are removed.

Next Crop:

- Avoid planting fir if needle rust incidence is high and ferns cannot be removed.

White fungal pustules are most often on the underside of needles.(B. Watt, Univ. of ME, Bugwood.org)

Infected balsam fir needles. (B. Watt, Univ. of ME, Bugwood.org)

Herbicide Injury

Herbicide Injury

Species Affected: All conifers

Importance: Herbicides that are improperly applied or that drift to nontarget plants while being applied can deform or kill needles, shoots, and sometimes entire trees.

Look For:

- *Yellow, bleached, or brown needles*, especially new needles on the side of the tree exposed to the herbicide.

- *Abnormal growth*; twisted needles; hooked, distorted, or swollen shoots.

- *Patterns of damage*, such as along one side of a field, along individual rows of trees, or only on one side of trees.

- **Pests that cause similar symptoms:** Air pollution, salt injury, drought, distorted needles from balsam twig aphid (true firs), or eriophyid mites

Biology: There are many different types of herbicides with different modes of action. They can enter trees through needles or through the roots. Hormone-type herbicides cause abnormal, exaggerated growth. Others slow growth by inhibiting photosynthesis or other life processes. The type and degree of injury will depend upon the herbicide applied, the concentration reaching the tree, the time of year, and the condition of the tree. Some products are lethal if applied when trees are actively growing but do not cause injury if applied when trees are dormant, such as in the late summer or fall. Other herbicides can be safely applied even during the active growth period.

Monitoring and Control: Check for injury during the first few weeks after herbicide application. Maintain a record of when and where applications occurred, the material applied, application rates, and information on weather conditions that occurred before and following applications.

- Remove dead shoots.

- Remove and chip or burn trees that are killed or severely affected by herbicides so that insect pests and pathogens cannot build up on them and spread to nearby healthy trees.

- Follow pesticide label directions carefully.

- Avoid applying herbicides directly to the foliage of trees.

- Reduce the chances of drift; do not apply on windy days and use the proper equipment.

Next Crop:

- Avoid planting Christmas trees near areas where herbicides are regularly used, such as powerlines, roadsides, and agricultural fields.

- Limit herbicide use and the number of applications whenever possible.

Hooked shoots damaged by picloram. (USFS - NCRS Archive)

Distorted red pine needles damaged by 2,4-D. (MN-DNR Archive, Bugwood.org)

Lirula Needlecast; Isthmiella Needlecast

Lirula nervata, Lirula mirabilis, Isthmiella faullii

Hosts: Balsam, Fraser, and white fir

Importance: These three fungi cause needlecast diseases of these fir species. Most injury occurs on small trees growing in cool, moist locations or on larger trees that are growing close to one another. Injury ranges from scattered brown needles to the loss of most 3- and 4-year-old needles. Over a period of years, repeated infection can reduce tree growth, cause bud and branch mortality on the lower portion of trees, and kill seedlings. Medium to high levels of disease decrease the quality of trees and result in boughs unusable for wreaths.

Look For:

JUNE TO JULY

- *Second-year needles in the lower crown that are pale green with patches of darker green, slowly turning brown.* Blister-like ridges develop on the upper surface of the brown needles. These ridges vary in color, shape, and location according to the specific species of fungus involved.

- *Third-year needles that are tan to brown with a dark line on the midrib of the underside of the needle.*

AUGUST TO SEPTEMBER

- *Third-year needles that are shades of brown or gray;*

needles may be cast or remain attached for several years.

Pests that cause similar symptoms: Rhizosphaera needle blight of fir, fall needle drop

Biology: The three fungi all have similar life cycles. Spores are released from infected needles during rainy periods in June, July, and August. These spores infect current-year needles. Newly infected needles remain symptomless until the following spring when they begin to discolor. Infected needles then become pale and patchy green and slowly turn brown. Spore-producing fruitbodies develop in the upper surface of these brown needles in late spring of the second year and mature during summer. In late summer of the second year, another type of spore-producing structure begins to form on the midrib on the lower surface of the needle.

Distinctive dark line on the underside of needles. (L. Haugen, USFS)

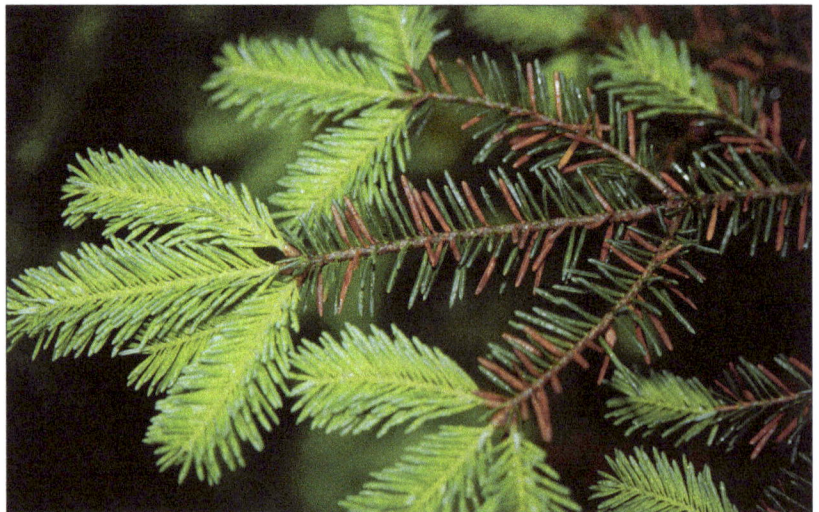

Discolored foliage. (S. Katovich, USFS, Bugwood.org)

By the summer of the third year, this structure looks like a dark line along the lower midrib. Infectious spores are released from this structure and the cycle begins again.

Monitoring and Control:
Examine trees of all ages at any time of the year. Look for second- and third-year needles that are various shades of brown and gray; these needles should also have one black line on the midrib of their lower surface.

- There is currently no fungicide registered to control these needlecasts.

- Promote good air movement by controlling weeds and pruning lower branches.

- Shear healthy trees first; disinfect tools after shearing.

- Do not shear during wet weather because spores could be carried to healthy trees on shearing tools.

Next Crop:

- Locate plantations in areas where there is good air drainage.

- Provide adequate space between trees to increase air movement around lower branches.

- Do not plant infected nursery stock.

- Do not introduce these pathogens into your plantation by transplanting infected native balsam or Fraser fir.

- Do not interplant fir in areas of your plantation where the disease is present.

Double ridge of pycnidia of Lirula mirabilis *on the upper surface of needles. (L. Haugen, USFS)*

Single ridge of pycnidia of Lirula nervata *on the upper surface of needles. (L. Haugen, USFS)*

Lophodermium Needlecast

Lophodermium seditiosum

Hosts: Scotch, Austrian, Eastern white, and red pine

Importance: This fungus kills red pine seedlings and causes dramatic browning of needles on Scotch pines of all ages. Severely affected trees, weakened by early needle loss, are unsuitable for sale as Christmas trees.

Look For:

MARCH TO APRIL

• *Brown spots with yellow margins on the needles.* Eventually needles turn yellow, then brown.

MAY TO JUNE

• *Brown needles, especially in the lower crown of the tree.* When severely affected, the entire tree will turn brown.

JUNE TO JULY

• *New shoot and needle growth on the tips of branches that hold dead, brown needles.* Most dead needles fall off in June, July, and August, leaving only tufts of new, green growth on the branch tips.

JULY TO OCTOBER

• *Black, football-shaped fruitbodies on dead needles* on the tree or on the ground. *Lophodermium* fruitbodies have a lengthwise slit down the middle and protrude from the needle when wet.

Pests that cause similar symptoms: Brown spot needle blight, pine needle scale, winter injury

Biology: Wind spreads spores from diseased needles to healthy needles in moist weather from August through October. The fungus overwinters in infected pine needles. In the spring, the fungus destroys the water-conducting system in needles, causing the foliage to turn brown.

Monitoring and Control: Examine trees of all ages in May or June. Look for needle spots and brown foliage on the lower branches of trees scattered throughout the plantation. If the

Black fruitbodies on dead needles. (A. Kunca, National Forest Centre - Slovakia, Bugwood.org)

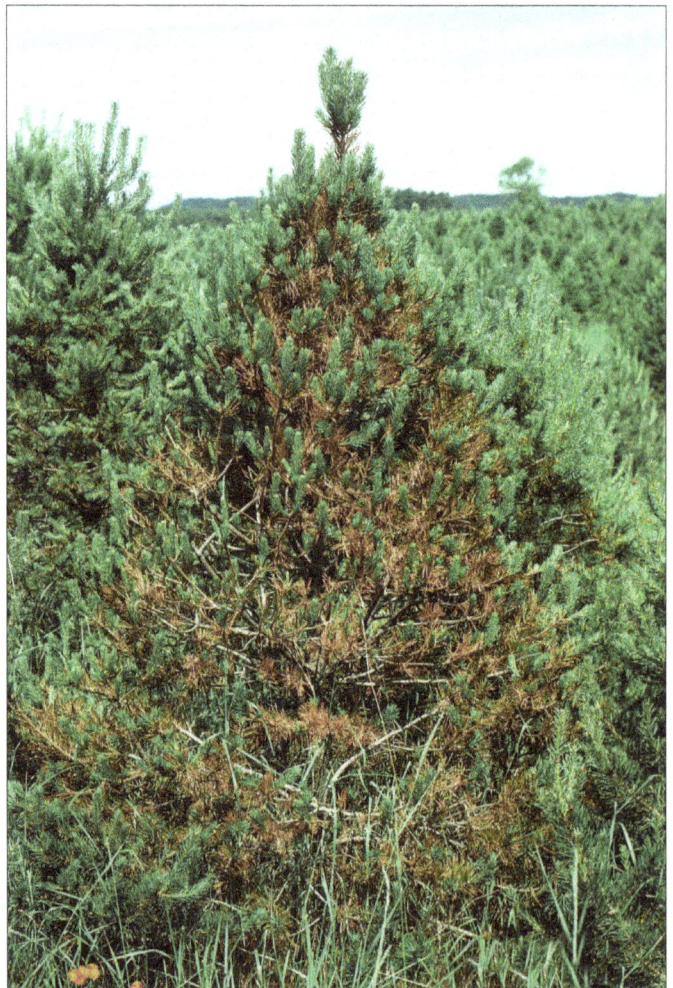

Brown needles on infected Scotch pine. (USFS - NCRS Archive, Bugwood.org)

incidence of disease is high, consider treating the entire plantation with a registered fungicide starting in late July.

- Irrigate seedlings in the morning so that they will have time to dry in the afternoon. This will avoid prolonged periods of moisture that favor infection.

- Promote good air movement by controlling weeds and pruning lower branches.

- Do not shear during wet weather because spores could be carried to healthy trees on shearing tools.

- Shear healthy trees first; disinfect tools after shearing.

- Do not leave live branches on stumps of harvested trees because the branches can serve as disease reservoirs.

- Apply a registered, preventive fungicide 3 or 4 times, once every 2 to 3 weeks, during the major infection period from late July through October. Apply more frequently if wet weather persists.

Next Crop:

- Plant only disease-free stock. If you suspect infection, have seedlings examined by a pest specialist.

- Do not plant seedlings next to windbreaks or abandoned plantations of the same species because the windbreaks or plantations can be disease reservoirs.

Football-shaped spore-producing fruitbody of Lophodermium. (USFS - NCRS Archive, Bugwood.org)

New shoot growth on damaged branch. (USFS - NCRS Archive)

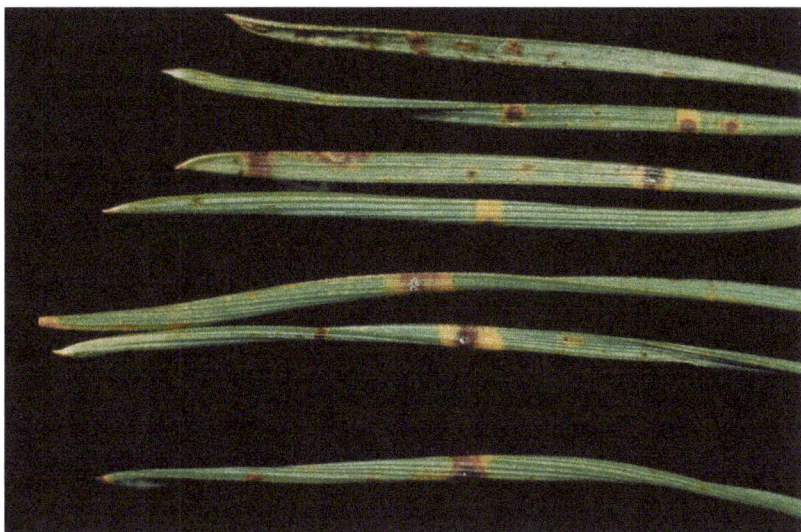

Brown spots with yellow margins on infected needles. (USFS Archive, Bugwood.org)

Pine Needle Rust

Coleosporium asterum

Hosts: Red, Scotch, and other two- and three-needle pines

Alternate Hosts: Goldenrod, aster

Importance: Most common on young trees, needle rust slows tree growth and causes unsightly foliage. When combined with insect pests and other agents that attack current-year foliage, needle rust may seriously damage or kill seedlings.

Look For (on pine):

APRIL TO MAY

- *Frosty-orange droplets at points of infection on needles at the onset of warm weather.*

MAY TO JUNE

- *Orange blisters* erupting from infected needles on lower branches.

Look For (on goldenrod or aster):

JULY TO AUGUST

- *Orange spores* on the undersides of leaves.

AUGUST TO SEPTEMBER

- *Orange, cushiony, bump-like structures* on the undersides of leaves.

Biology: This fungus needs both pine and aster or goldenrod to complete its 1-year life cycle consisting of five spore stages. Rust spores produced on pine needles do not infect pines. Windborne spores from pine needles infect goldenrod or aster, and only spores produced

Orange fungal blisters erupting from needles. (R. Anderson, USFS, Bugwood.org)

on these alternate hosts can infect pines. The rust fungus overwinters in pine needles and is perennial, so it can survive 2 consecutive years of unfavorable weather.

Monitoring and Control: Examine 3- to 6-year-old trees in May and June. Check needles on the lower branches. If you find a high incidence of orange blisters and serious foliage loss on these trees, remove goldenrod and aster in and around the plantation before August. Without its alternate host, the fungus will not be able to complete its life cycle and infect pines.

- Mow goldenrod and aster. These plants are perennial and will need mowing each year until the trees are old enough for the rust to have little or no impact on tree quality.

- Kill goldenrod or asters within 1,000 feet of newly planted seedlings before August by applying a registered herbicide.

Next Crop:

- Remove tall grass, weeds, goldenrod, and aster in and around the plantation before planting.

- Avoid planting on humid sites that receive limited sunlight.

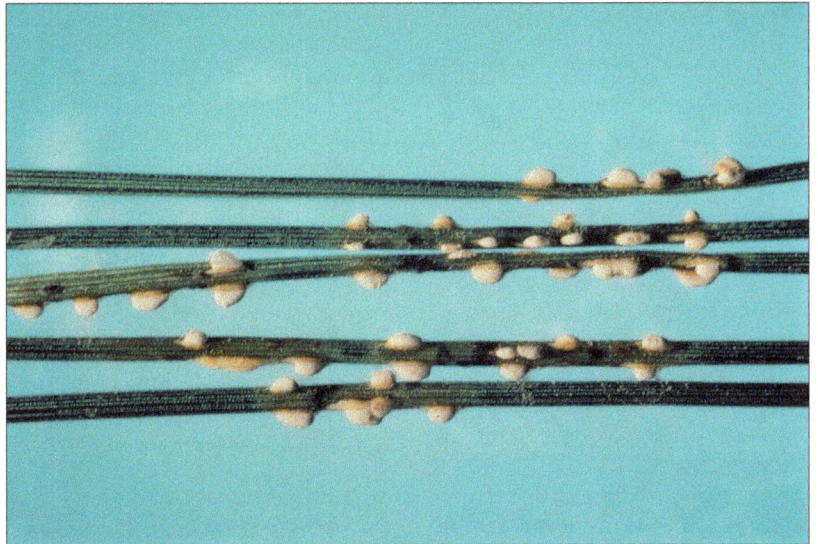

Orange fungal blisters erupting from needles. (USFS - NCRS Archive)

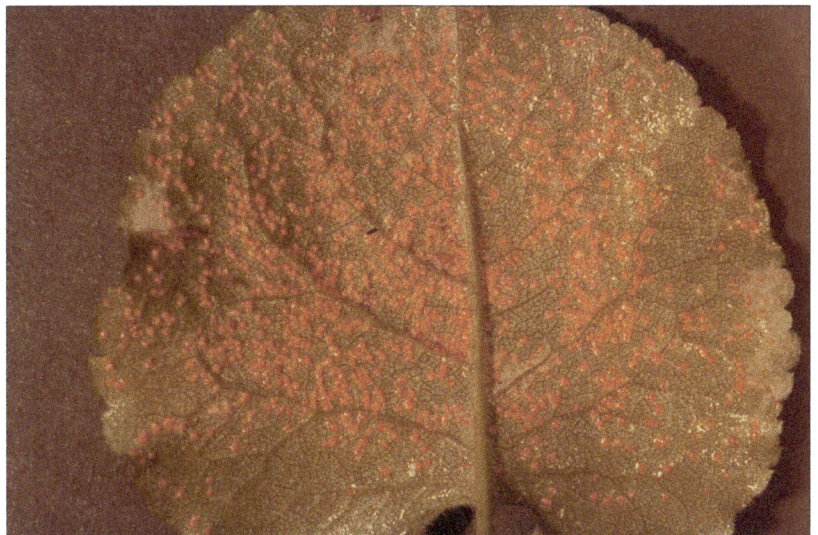

Orange, cushiony bumps on the underside of goldenrod or aster leaves. (USFS - NCRS Archive)

Pine Needle Scale

Chionaspis pinifoliae

Hosts: All pines and spruces, Douglas-fir, eastern redcedar, Fraser fir

Importance: This insect feeds by sucking sap from needles. At high densities, feeding can reduce tree growth and vigor. Needles on heavily infested trees may be nearly covered by the white scales. Foliage may be sparse and discolored, reducing the grade and value of Christmas trees. If infestations persist, shoots or entire branches can be killed.

Look For:

- *Many white or light-yellow, oyster-shaped scales,* about 1/10 inch long, covering the needles.
- *Trees that appear grayish green*; high scale densities can discolor needles, causing trees to turn off-color.

Pests that cause similar symptoms: Some needlecast diseases, sawflies (eggs look like scales)

Biology: Small, reddish eggs overwinter on the needles beneath the hard, white covering of dead female scales. The first generation of crawlers (nymphs) hatch in May or June. Crawlers are mobile and move about until they settle on a needle and begin to suck sap. As the pink crawlers feed, they change to an opaque, yellowish color, called the hyaline stage, and then begin to secrete the hard, white wax that will cover their body. Scales mature in early July and produce a second generation of eggs that hatch in July or early August. Those scales eventually lay the eggs that overwinter.

Monitoring and Control:

Inspect trees of all ages, looking for white flecks on the needles before lilacs bloom in spring. Be sure to check needles on lower branches where many infestations begin.

- Cut, remove, and destroy severely infested trees.
- Many predatory lady beetles feed on pine needle scale. Jagged tears in the hard, white armor of scales are evidence of lady beetle predation. Tiny parasitoid wasps also kill scales. These wasps leave a small, round hole in the dead scale. If you see signs of predation or parasitism, control measures are probably not needed.
- Scout trees regularly. A hand lens can be helpful when monitoring egg hatch and scale development. Spot treat individual trees if 5 to 10 white flecks per shoot are observed.
- See table 1 (page 22) for degree day information.
- If infested trees are abundant, spray trees thoroughly with a registered insecticide or horticultural oil to control first-generation crawlers. The best time to spray is when most scale eggs have hatched and immature scales are in the hyaline stage.
- Spray trees again if needed in late July or early August to control the second generation of crawlers. Timing the second spray is more difficult because eggs may be hatching over a 2- to 3-week period. The ideal timing for the second spray is usually around 1500 DD_{50}.

Next Crop:

- Reduce spraying for pine needle scale and other pests in new plantings whenever practical. Scales often reproduce rapidly after insecticide sprays because the insecticides kill the natural enemies of the scales.

Closeup of pine needle scales. (USFS - NCRS Archive)

Heavily infested needles. (USFS - NCRS Archive)

Pine Thrips

Gnophothrips spp.

Hosts: Scotch and Austrian pines

Importance: When severe, pine thrips feeding can distort needles and weaken, stunt, or kill Christmas trees or seedlings. Severely injured nursery seedlings are unfit for outplanting, and injured trees are unsuitable for Christmas tree sale.

Look For:

- *Discolored, crooked needles, particularly on the upper branches.* Severely injured trees die and lose their needles.

- *Curled needles* anywhere on the trees. Needles growing from the same sheath may differ in size.

- *Brownish wounds, ⅛ to ¼ inch wide, on the needles.*

LATE APRIL TO OCTOBER

- *Orange-yellow or black insects,* up to $1/16$ inch long, on the buds or new needles. Use a hand lens to see them clearly.

Biology: This thrips overwinters in the soil. The winged, black adult thrips lay their eggs in early spring at about the time of bud break. Several subsequent generations produce thousands of individuals that feed on trees throughout the summer. Hot, dry weather favors their buildup.

Monitoring and Control:
Examine trees in late summer. If an average of 10 percent of the needles in the tops of the trees shows damage, treat the entire nursery or plantation the following spring. Treated trees will usually outgrow the injury in 2 to 3 years.

- Irrigate seedlings and young trees frequently during hot, dry weather. Water with an overhead sprinkler system early in the morning to discourage thrips and to reduce the likelihood of needlecast disease.

- Do not ship infested nursery stock or infested trees because overwintering thrips hitchhike to new areas this way.

- Thoroughly spray trees with a registered insecticide once in very early spring before eggs are laid to control the first generation of emerging adults. If you delay treatment until later in the season, two or three applications may be needed for complete control.

Next Crop:

- Do not bring infested transplant stock into fields.

Brownish wound and a black adult thrips. (D. Mosher, MI DNR)

Ploioderma Needlecast

Ploioderma lethale

Hosts: Austrian, red, and other two- or three-needle pines

Importance: This fungus causes needle browning and defoliation. Severely affected trees can become stunted or killed after repeated years of infection.

Look For:

FALL

- *Yellow to brown spots and bands* on mature needles.

SPRING and SUMMER

- *Straw-brown needles* beyond coalesced spots; lower portions of needles may remain green.

- *Elongated, black fungal fruitbodies* in dead portions of attached live needles and cast dead needles.

- *Casting* of affected needles that leaves only the new foliage.

Pests that cause similar symptoms: Pine needle scale, winter injury, Dothistroma needle blight

Biology: Spores produced in fruitbodies on attached and fallen infected needles are dispersed in spring, infecting elongating needles. Symptoms develop in the fall and the fungus overwinters in the diseased needles, completing the 1-year disease cycle.

Monitoring and Control:
Symptoms are first noticed in the fall.

- Promote good air movement by controlling weeds and pruning lower branches.

- Remove severely affected lower branches to reduce sources of the fungus.

- Do not leave live branches on stumps of harvested trees because the branches can serve as disease reservoirs.

Next Crop:

- Plant only disease-free nursery stock.

- Inspect trees carefully for the first few years after planting for disease symptoms.

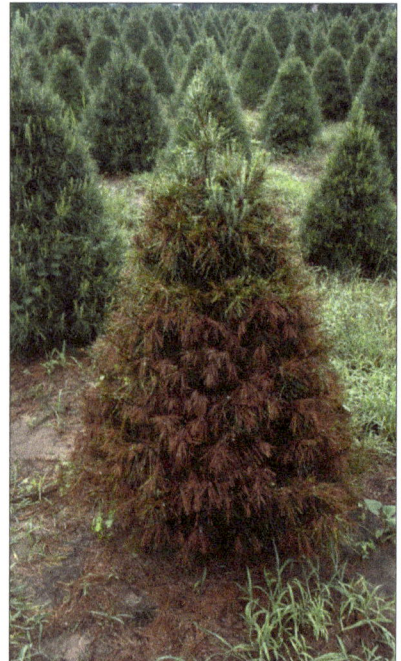

Severely affected tree with casting needles. (D.J. Moorhead, Univ. of GA, Bugwood.org)

Straw-brown needle spots with black, elongated fruitbodies. (G. Hudler, Cornell Univ.)

Rhabdocline Needlecast

Rhabdocline Needlecast

Rhabdocline pseudotsugae

Host: Douglas-fir, especially the Rocky Mountain variety

Importance: The browning and early needle loss caused by this fungus makes affected Douglas-firs unsalable as Christmas trees.

Look For:

LATE SUMMER

- *Yellow spots on infected, current-year needles*, especially in the lower portions of trees. Yellow spots enlarge and needles appear mottled.

EARLY SPRING

- *Yellowish-brown to purplish-brown needles.*

EARLY SUMMER

- *Shedding of brown needles.* Severely diseased trees will retain only their current-year needles.

Pests that cause similar symptoms: Pine needle scale, Swiss needlecast

Biology: Fruitbodies that develop on the brown needles release spores during moist weather from May to July. Windborne spores infect only young needles during shoot elongation.

Monitoring and Control: Inspect 5- to 10-year-old trees in May. Examine the 2-year-old needles. If you find a high incidence of disease, consider treating the entire plantation in early spring.

- Promote good air movement by controlling weeds and pruning lower branches.
- Do not shear during wet weather because spores could be carried to healthy trees on shearing tools.
- Shear healthy trees first; disinfect tools after shearing.
- Do not leave live branches on stumps of harvested trees because the branches can serve as disease reservoirs.
- Remove diseased lower branches and severely affected trees early in the rotation to prevent disease buildup.
- Apply registered fungicides when buds burst; repeat every 7 to 10 days until the buds are fully open. Fungicides applied after buds are fully open will not be effective. Fungicide treatments used for Rhabdocline needlecast will also control Swiss needlecast.

Next Crop:

- Plant only disease-free nursery stock.
- Plant disease-resistant varieties of Douglas-fir. If you plant a Rocky Mountain variety, select seed sources that show resistance to Rhabdocline needlecast.

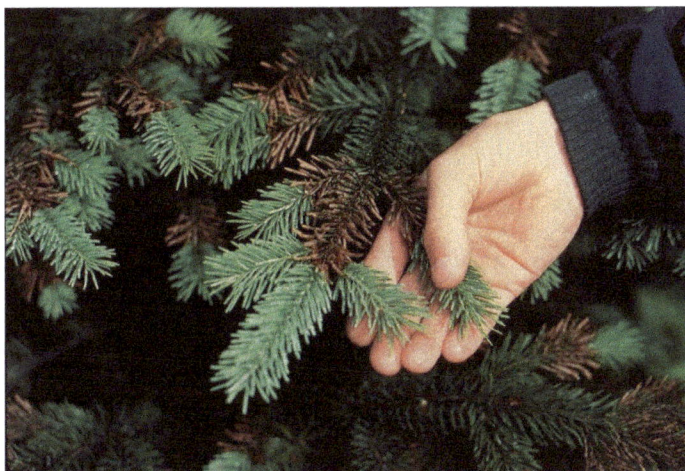

Reddish-brown, diseased needles. (USFS - NCRS Archive)

Needle spots and mottling of infected needles. (J.W. Schwandt, USFS, Bugwood.org)

Rhizosphaera Needle Blight of Firs

Rhizosphaera pini

Hosts: Balsam, Fraser, and other firs

Importance: This fungus can infect foliage of all ages and causes needle droop, discoloration, and death, reducing the quality and value of trees. Greatest disease damage is to needles on lower branches where it may result in branch death. Severe injury occurs on stressed trees or trees growing in shaded, damp, and cool areas.

Look For:
- *Yellow to tan needles.* Needles turn grayish tan and die. Needles may also droop. Before new growth occurs in the spring, all needles on individual severely affected branches may be dead from the tip of the branch to the tree trunk.

- *Tiny, black fruitbodies on the undersides of green, yellow, and grayish-tan needles* that can be seen with a hand lens. The fruitbodies emerge from the needle stomata (tiny, pore-like openings on the underside of needles) and often have a speck of white wax on top of them.

Pests that cause similar symptoms: Lirula needlecast

Biology: Symptoms appear after periods of rainy weather and cool temperatures at any time during the growing season. Spore release and infection probably occur throughout the growing season during favorable environmental conditions. Needles of any age can be infected. Needles may discolor soon after infection before fruitbodies develop or fruitbodies can develop while needles are still green. Many infected needles remain on the tree over winter and into the next summer.

Monitoring and Control: Examine trees and foliage of all ages throughout the year. The previous year's damage is most evident early in the spring before new growth develops, but symptoms and damage can develop anytime during the growing season. Fungicide recommendations have not been developed. The best control is cultural management to avoid stress and conditions favorable to the fungus.

- Promote good air movement by controlling weeds and pruning lower branches.

- Do not shear during wet weather because spores could be carried to healthy trees on shearing tools.

- Disinfect tools after shearing. Shear healthy trees first.

- If disease is localized, remove and burn infected branches and trees. This should reduce, but may not eliminate, future infections.

- Do not leave live branches on stumps of harvested trees because the branches can serve as disease reservoirs.

Next Crop:
- Plant trees with adequate space between them to provide good air movement.

- Do not grow fir in shady areas or where cool, moist air collects.

- Plant only healthy stock, and do not interplant fir seedlings in fields where older, diseased trees are present.

Small, black fruitbodies emerging from stomata of dead needles. (R.S. Kelley, VT Dept. of Forests, Parks and Recreation, Bugwood.org)

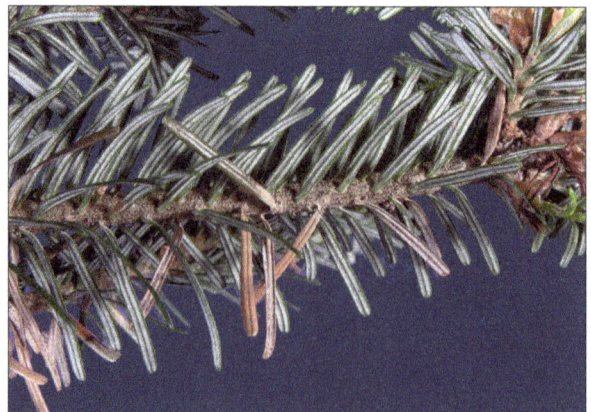

Scattered grayish-tan, dead needles remaining on a tree. (R.S. Kelley, VT Dept. of Forests, Parks and Recreation, Bugwood.org)

Rhizosphaera Needlecast

Rhizosphaera Needlecast of Spruce

Rhizosphaera kalkhoffii

Hosts: Colorado blue spruce, occasionally white spruce

Importance: This fungus causes needles to turn purplish brown and fall prematurely, thereby reducing the vigor and market value of trees grown for Christmas trees. Three or four years of early needle loss kills branches and, in severe cases, the entire tree. Disease damage is most severe on branches in the lower crown and later spreads upwards.

Look For:

LATE FALL OR EARLY SPRING

* *Fuzzy, black fruitbodies* of the fungus protruding out of tiny, pore-like openings (stomata) on both green and yellow needles. Use a hand lens to more easily see them. The yellow needles later turn purplish brown.

JULY TO AUGUST

* *Purplish-brown, 1- and 2-year-old needles*, most commonly on the lower branches. Most of these needles drop off by late fall.

Pests that cause similar symptoms: Stigmina needlecast caused by *Stigmina lautii*, sudden needle drop caused by *Setomelanomma holmii*, drought, pine needle scale, spruce spider mite, Cytospora canker

Biology: Some infected needles remain on the tree throughout winter. The next spring, spores from these infected needles are

Fuzzy, black fruitbodies in stomata on a needle. (MN DNR Archive, Bugwood.org)

Purplish-brown 1- and 2-year-old needles. (USFS - NCRS Archive, Bugwood.org)

rain splashed or manually spread by equipment to newly emerging needles. Although infection is possible from mid-April to October, it usually occurs during wet weather shortly after bud break.

Monitoring and Control: Inspect trees of all ages during May. Examine the white rows of stomata on the 2-year-old needles on bottom branches with a hand lens. If a high incidence of the branches have needlecast fruitbodies on the needles (*R. kalkhoffii, Stigmina lautii*) or on the branches (*Setomelanomma holmii*), consider treating the entire plantation in spring and summer. To determine the best control strategy, it is important to correctly identify the pathogen causing damage.

- Promote good air movement by controlling weeds and pruning lower branches.

- Do not leave live branches on stumps of harvested trees because the stumps can serve as disease reservoirs.

- Do not shear during wet weather because spores could be carried to healthy trees on shearing tools.

- Shear healthy trees first; disinfect tools after shearing.

- Apply a registered, preventive fungicide in spring when needles are half elongated and again when needles are fully elongated. Two years of treatment should permit most trees to develop full foliage. If treated early, Rhizosphaera needlecast can be controlled in 1 year.

- Cultural control strategies for Rhizosphaera needlecast of spruce will also work for look-alike Stigmina needlecast and sudden needle drop, but fungicides might not be effective. Sudden needle drop is associated with poor site conditions.

Next Crop:

- Plant only disease-free stock.

- Maintain high tree vigor.

Colorado blue spruce with diseased and dead lower branches. (USFS - NCRS Archive, Bugwood.org)

Salt Injury

Species Affected: Eastern white pine, red pine, and balsam fir are very susceptible; most conifers can be damaged on occasion

Importance: Salt injury of many conifer species causes browning and early needle loss, thereby degrading or making Christmas trees unsalable. Severe injury can kill branches and small trees. Trees affected by salt are usually those growing next to major roads and intersections. Damaged trees are generally found only within one or two outer rows in a plantation, usually within 100 feet of roads. Salt injury can impair cold hardiness.

Look For:

APRIL TO JUNE

- *Browning of needles* on the side of trees facing the road. These needles generally drop and new buds usually develop normally. For affected needles, there is a clear demarcation between healthy and brown, damaged tissue.

Pests that cause similar symptoms: Winter injury on most conifers, brown spot needle blight, Dothistroma needlecast, Lophodermium and Cyclaneusma needlecasts

Biology: Damage results from salt spray onto needles, buds, and twigs. The browning is often concentrated on the side of the tree facing major roads. Trees that grow where salt accumulates, such as small depressions that drain water from roads or ditches along roads, can also be damaged by salt uptake through roots. The affected foliage falls off during the spring and early summer, thinning the crown. New growth will make the tree appear otherwise healthy. However, trees repeatedly affected by salt often become stunted and grow slowly. They may eventually be killed by prolonged exposure. Weakened trees may be killed by insect pests or disease.

Monitoring and Control: Monitor plantations between April and June. Look along the edges of plantations that are adjacent to major roads. The amount of injury can vary greatly between years, depending upon the amount of salt used on roads during the winter.

- Harvest trees as soon as possible after an injury-free winter.

Next Crop:

- Avoid planting susceptible species, such as white pine, red pine, and balsam fir, along major roads.

- Plant resistant species. Scotch pine, Colorado blue spruce, and Black Hills spruce are relatively tolerant of salt.

Left: Salt injury on eastern white pine. (USFS - NCRS Archive)

Below: Browning of damaged needles. (J. O'Brien, USFS)

Spruce Needle Rusts

Chrysomyxa spp.

Hosts: Black, white, and Colorado blue spruce; occasionally Norway spruce

Alternate Hosts: Labrador tea, leather leaf (Weir's spruce cushion rust caused by *C. weirii* does not need an alternate host)

Importance: During epidemics, infected trees will lose 25 to 75 percent of their new needles, leaving them unsuitable for sale. Repeated years of high disease severity will slow tree growth, but will rarely kill trees.

Look For:
JULY TO AUGUST

- *Yellow, current-year needles* anywhere on the tree.
- *Whitish blisters filled with yellow spores* on the undersides of current-year needles.

AUGUST TO SEPTEMBER

- *Shedding of infected needles.*

Biology: The species of fungi that cause spruce needle rusts need an alternate host to complete their life cycles. During the summer, windborne spores released from fungal blisters on spruce needles infect plants such as Labrador tea or leather leaf. The fungi overwinter on these alternate hosts, and spores released from them reinfect spruce the following spring. The only exception is *C. weirii*, which overwinters on needles infected the previous season and whose spores are blown by wind or rain splashed to developing needles the following year where infection takes place.

Monitoring and Control:
Inspect trees in late summer for diseased needles.

- No control is usually necessary because severe disease incidence rarely occurs in consecutive years.

- Provide good air movement among trees to reduce conditions favorable for disease development.

- If possible, remove alternate hosts near plantations to reduce disease incidence.

- Consult a pest specialist to determine the rust species present. Cultural and chemical control options are available for needle rust caused by *C. weirii*.

Next Crop:

- Avoid planting spruce near bogs or wetlands that contain Labrador tea and leather leaf.

- Plant resistant species of spruce, such as Norway or Black Hills. White spruce is moderately resistant, but black and Colorado blue spruce are extremely susceptible.

Clockwise from far left: Severely affected tree (S. Katovich, USFS); Whitish blisters filled with yellow spores on infected needles (USFS - NCRS Archive); Infected, discolored needles. (USFS - NCRS Archive)

Spruce Spider Mite

Oligonychus ununguis

Hosts: All Christmas tree species

Importance: Mites may discolor, degrade, or kill Christmas trees of all ages. Injury is most common during prolonged dry periods, on droughty soils, and where overuse of insecticides has killed the natural enemies of the mites.

Look For:

- *Yellowish to rusty-brown shoots.* Look closely to see yellow mottling on needles.

- *Fine webbing between the needles, especially on older needles near the stem*. This may require a hand lens.

- *Dark-green to brown mites*, less than 1/50 inch long, on needles or webbing. To see mites, shake an injured branch over a piece of white paper and use a hand lens to see the tiny, moving specks.

Pests that cause similar symptoms: Admes mite, Rhizosphaera needlecast, drought

Biology: Mite eggs overwinter on shoots. Nymphs hatch in May or June and feed on sap in the needles. The feeding mites produce silk webbing. Adults appear in June or early July; three or more generations follow at 2½- to 3-week intervals until the weather turns cold. Mites can be windblown to new areas or carried on infested nursery stock. Mites will go dormant during hot weather if the daily maximum temperature is above 85 °F for more than 5 days. They will resume activity when the weather cools.

Monitoring and Control: Examine trees of all ages throughout the growing season, beginning in June. Mites usually begin feeding on older needles near the trunk, but can feed on current-year needles if populations build to high densities. Delay control if injury or webbing is barely noticeable or if rainfall/humidity is high. However, if injury occurs during dry weather or if trees are to be harvested that year, treat individual infested trees as soon as you notice symptoms.

- See table 1 (page 22) for degree day information.

- Spray infested trees with a registered miticide in early June to early July when you first see mite activity. Some miticide products should be applied only once a year, ideally in early summer. Use miticides that do not kill beneficial insects and predatory mites that will eat spider mites. Other miticide products may need to be applied at 2-week intervals to kill mites emerging from eggs.

- Alternatively, spray trees thoroughly with a dormant oil in early spring before tree growth starts.

Next Crop:

- Avoid planting on droughty soils, especially when planting spruce or fir.

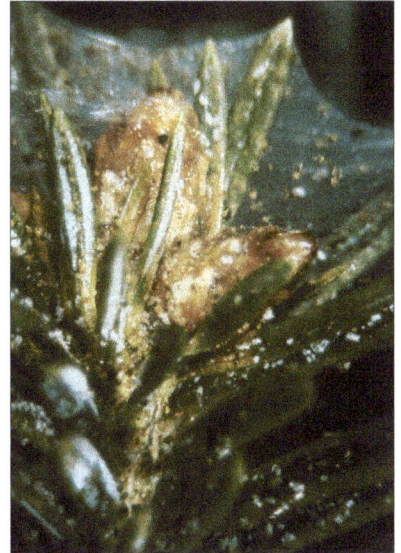

Fine webbing between needles. (USFS - Region 4 - Intermountain Archive, Bugwood.org)

Spruce spider mite. (USFS Archive, Bugwood.org)

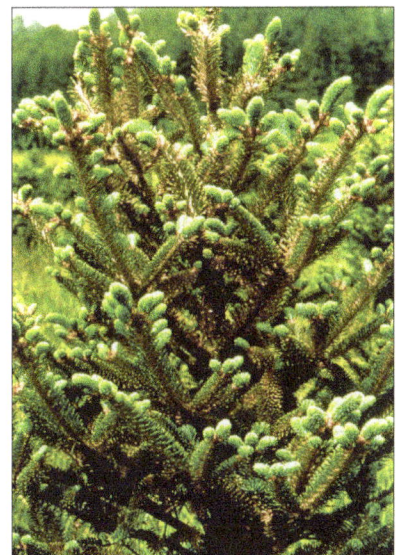

Fir with rust-colored needles. (USFS - NA Archive, Bugwood.org)

Swiss Needlecast

Phaeocryptopus gäeumannii

Host: Douglas-fir

Importance: The browning and early needle loss caused by this fungus result in trees that are unsuitable for sale as Christmas trees.

Look For:

SPRING AND FALL

- *Rows of fuzzy, black fruitbodies* in tiny, pore-like openings (stomata) on the undersides of both green and yellow needles. Use a hand lens to see them. Yellow needles later turn brown.

JULY TO AUGUST

- *Brown, 2- and 3-year-old needles*, especially on the lower branches. These needles fall off in late fall and winter.

Pests that cause similar symptoms: Pine needle scale, Rhabdocline needlecast

Biology: Airborne spores infect needles on new shoots during wet weather at the time of bud break. The fungus is commonly spread on infected nursery stock.

Monitoring and Control: Inspect trees during May. Examine the white rows of stomata on the 2-year-old needles with a hand lens. If a high percentage of needles have fruitbodies, consider treating the entire plantation before summer.

- Promote good air movement by controlling weeds and pruning lower branches.

- Do not shear during wet weather because spores could be carried to healthy trees on shearing tools.

- Disinfect tools after shearing. Shear healthy trees first.

- If disease is localized, remove and burn infected branches and trees. This should reduce, but may not eliminate, future infections.

- Do not leave live branches on stumps of harvested trees because the branches can serve as disease reservoirs.

- Apply a registered, preventive fungicide in spring when the new shoots are ½ to 2 inches long. Apply again in 2 to 3 weeks and once again if rainfall is abnormally high. Two years of treatment should restore most trees to full foliage. Treat nursery stock every 2 weeks from bud break to mid-August.

Next Crop:

- Inspect planting stock carefully. Plant only disease-free stock.

- Plant stock grown from Pacific Coast sources because these trees have exhibited disease resistance.

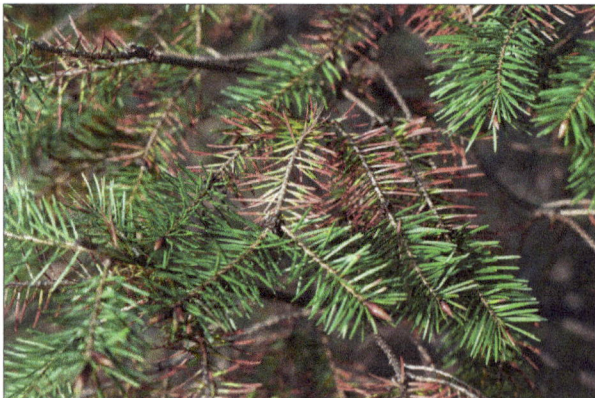

Brown 2- and 3-year-old infected needles. (USFS - NCRS Archive, Bugwood.org)

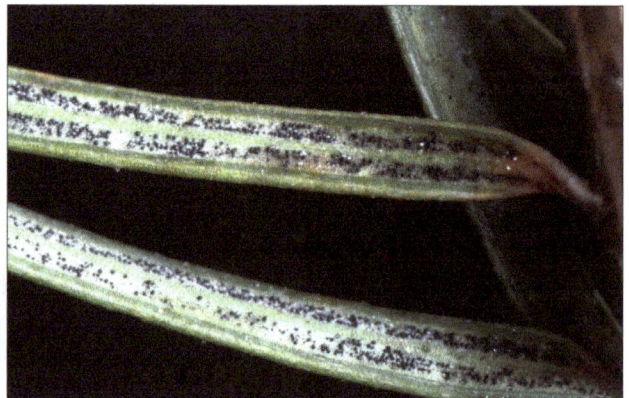

Fuzzy black fruitbodies in needle stomata. (P. Kapitola, State Phytosanitary Administration, Bugwood.org)

Winter Injury

Species Affected: Short-needled Scotch pine varieties; Austrian and white pine; Norway spruce; balsam, white, and Fraser fir; occasionally other Christmas tree species

Importance: Winter burn and winter drying cause needles to turn brown and fall off, thereby degrading or making Christmas trees unsalable. Severe injury for several years may kill branches and occasionally kill trees.

Look For:

- *Lack of foliage*, especially on the south side of trees where injury is usually most severe.

APRIL TO JUNE

- *Browning of entire tree or brown needles* above the previous snowline at the onset of warm weather. These needles drop and new buds usually develop normally.

Pests that cause similar symptoms: Brown spot needle blight on Scotch pine, Dothistroma needlecast on Austrian pine, Lophodermium and Cyclaneusma needlecast on Scotch pine, salt injury

Biology: Water cannot move easily in soil and in trees in the winter. So when needles lose moisture in the winter, it cannot always be replaced. This is called winter drying. It occurs most often when the soil around tree roots is frozen, and warm winds and bright sun dry out and damage needle and bud tissues. A second type of winter injury is called winter burn. Winter burn occurs when needles are damaged after rapid temperature

Damaged trees; note browning of trees above snowline. (USFS - NCRS Archive, Bugwood.org)

changes, particularly on the south side of trees where exposure to the sun is greatest. Rapid temperature drops can occur at sunset and sunrise or when sunlight is suddenly blocked by other trees, hills, or buildings.

Sometimes winter burn and winter drying will occur together, occasionally complicated by drought. The amount of injury can vary greatly from year to year depending on weather events. In addition, certain tree species and varieties are much more susceptible than others. Exotic trees (those grown outside of their native ranges) tend to be especially susceptible to winter injury.

Monitoring and Control: Look for browning between April and June. Keep accurate records of winter injury throughout the life of the trees.

- Harvest old, susceptible trees as soon as possible after an injury-free winter.

- Figure the cost-benefit of keeping young, susceptible trees. If not profitable, destroy the trees and replant.

Next Crop:

- Plant resistant species and varieties, such as the long-needled Scotch pine varieties; red pine; and white, blue, and Black Hills spruce.

- Avoid planting susceptible species and varieties, such as Spanish Scotch pine, Austrian pine, or Norway spruce. If you do plant susceptible trees, plant them in areas that are protected from the wind.

Needle Feeding

Portions of needles or entire needles are chewed off. Several needles may be clipped and webbed together into tubes, bags, or nests. Search for pellets of waste on the ground beneath injury to verify that damage is due to feeding and not simply needle drop.

Bagworm

Thyridopteryx ephemeraeformis

Hosts: Eastern redcedar, spruce, fir, eastern white pine, Scotch pine

Importance: Bagworm feeding results in trees with thin foliage and brown cases attached to twigs. Damaged trees are unfit for Christmas tree sale and may be killed if larvae strip off all the foliage. This insect is most common from southern Minnesota south to Missouri and then east through Illinois, Indiana, Ohio, and southern Pennsylvania.

Look For:

- *Sparse or stripped foliage*, especially at the top of the tree. Shoot tips may be flagged (discolored and deformed).

- *Conspicuous, brown silk bags*, 1¾ to 2 inches long, with embedded needle particles.

Biology: The wingless female moths lay eggs in August and September inside silken bags. In late May newly hatched larvae (caterpillars) emerge from bags on silken strands that are readily blown in the wind to nearby trees. Once they have landed on a tree, the larvae feed on needles and spin silken bags they cover with bits of needles. Larvae carry these bags around as they feed on needles all summer. Once they have completed development, they secure the bags to twigs. Winged adult males fly to mate with females, who remain in their bags where they lay the eggs that will spend the winter on the tree. Because caterpillars move only short distances and females never leave the securely attached bag, damage may be confined to a single tree and its nearest neighbors.

Monitoring and Control: Begin monitoring trees of all ages in May and continue throughout the growing season. Treat by hand unless infestation is severe or widespread.

- Handpick and destroy silken bags, which will contain eggs, moths, or caterpillars, depending on the time of the year. Bags can be destroyed by soaking them in a bucket of soapy water or by burning.

- Cut out and destroy individual trees that are severely infested.

- For widespread infestations, spray trees with a registered insecticide. Treat when caterpillars are small and before much feeding injury has occurred.

Next Crop:

- Before planting, remove infested trees in natural stands or windbreaks near the site.

Individual bag. (J. Hanson, USFS, Bugwood.org)

Bagworm infestation. (PA DCNR - Forestry Archive, Bugwood.org)

Balsam Fir Sawfly

Neodiprion abietis

Hosts: Balsam and Fraser fir, white and black spruce

Importance: This sawfly is not commonly encountered, but when it is present, populations can build to high densities capable of damaging trees. Larvae feed on old needles and avoid current-year needles. Feeding results in thin, transparent crowns with few interior needles.

Look For:

SEPTEMBER TO MARCH

- *Thin crowns;* older interior needles will be partially eaten and often turn reddish brown. Empty cocoons and distinctive egg slits in needles will confirm sawfly activity.

JUNE TO AUGUST

- *Reddish or brown needles in July,* in the interior of the crown on twigs that are 1 year old and older.

- *Sawfly larvae that are dark green with darker longitudinal stripes and a dark head.* When mature, they are about a half-inch long.

- *Clusters of young larvae in early or midsummer.* Larvae will disperse as they mature and grow.

Biology: Larvae hatch in early summer and begin feeding on the previous year's needles. Larvae avoid current-year needles unless populations are very high. Young larvae feed in colonies of 30 to 100. As larvae grow and develop, they disperse and feed singly on old needles. Young larvae consume only parts of needles. Older larvae will consume entire needles, except for the central vein or midrib. Damaged needles wither and dry out, then eventually drop to the ground. Larvae feed for roughly 4 weeks and then pupate in cocoons that are attached to needles. Adults emerge in late summer or fall and lay eggs on host trees. Winter is spent in the egg stage.

Monitoring and Control:
Inspect plantations beginning in early summer. Look for trees with red-brown interior needles. If control is needed, treat infested trees as soon as you see larvae and make sure to get good coverage.

- Clip off and destroy colonies of young larvae before extensive defoliation occurs.

- Spray infested trees with a registered insecticide if small larvae are abundant.

Egg slits in needles look like spots. (G. Leclair, CFS, Atlantic For. Centre)

Group of balsam fir sawfly larvae. (G. Leclair, CFS, Atlantic For. Centre)

European Pine Sawfly

Neodiprion sertifer

Hosts: Scotch, Austrian, and red pine

Importance: In early spring, larvae of this sawfly strip the old needles from pine Christmas trees, giving them a thin look in autumn. Trees can outgrow the injury and recover their full, dense foliage in 2 or 3 years.

Look For:
* *Sparse or missing old foliage* anywhere on the tree.

APRIL TO MAY
* *Tufts of dry, straw-like needles* that remain behind emerging, new green growth.

APRIL TO JUNE
* *Green-striped larvae* up to 1 inch long, with shiny, black heads, in clusters on the old foliage. There may be as many as 100 larvae in a cluster.

SEPTEMBER TO APRIL
* *Rows of yellow spots on needles*; these are where eggs have been laid in slits in the needles.

Biology: Adult females lay their eggs in needles in September and October. The eggs overwinter, and hatching begins in mid-April to early May. Young larvae feed in groups on the outer edges of old needles and produce tufts of dead needles. Older larvae eat entire needles and leave only the needle sheaths. One larval colony of 80 to 100 larvae can eat all the old foliage off a 2-foot-tall tree; 15 to 20 colonies can completely defoliate a larger tree. When full grown, larvae drop to the ground, spin cocoons, and pupate. There is only one generation a year.

Monitoring and Control: Begin inspecting trees 3 years before harvest or when many small trees are repeatedly defoliated. Look for eggs from September to April, and look for larvae or damage from April to June. Treat infested trees as soon as you see larvae.

* See table 1 (page 22) for degree day information.

* If colonies are few and scattered, knock the larvae off and crush them.

European pine sawfly larvae eat older needles in early spring. (S. Katovich, USFS, Bugwood.org)

- Growers can make a virus suspension from diseased sawfly colonies that can be applied much like an insecticide. Look for freshly killed, diseased larvae that are soft, black, and hanging head down. Place 100 to 150 dead larvae in a pint of chlorine-free water (distilled or rainwater), and allow them to disintegrate over the summer. Filter the solution through a fine cotton cloth to remove any debris. The stock solution is then ready to use the following spring. Prepare a spray solution with 3 teaspoons of stock solution per 6 gallons of water. Add 5 teaspoons of powdered milk or some other "sticker" to each gallon. Drench larvae shortly after they emerge from the eggs. The virus will usually kill larvae in 4 to 10 days.

- Alternatively, apply a horticultural oil, insecticidal soap, or registered insecticide that lists sawflies on the label.

- Do not ship nursery stock or Christmas trees that have sawfly eggs in the needles. Larvae may hatch in buyers' homes.

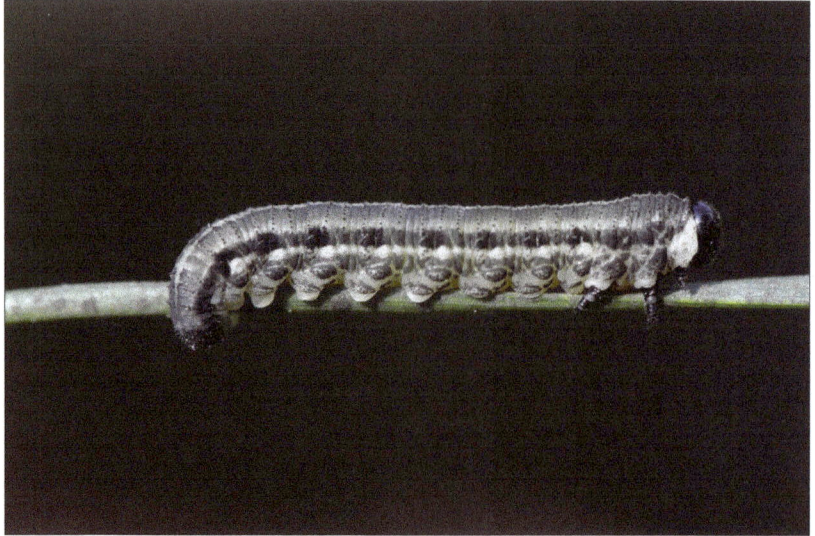

European pine sawfly larva. (S. Katovich, USFS, Bugwood.org)

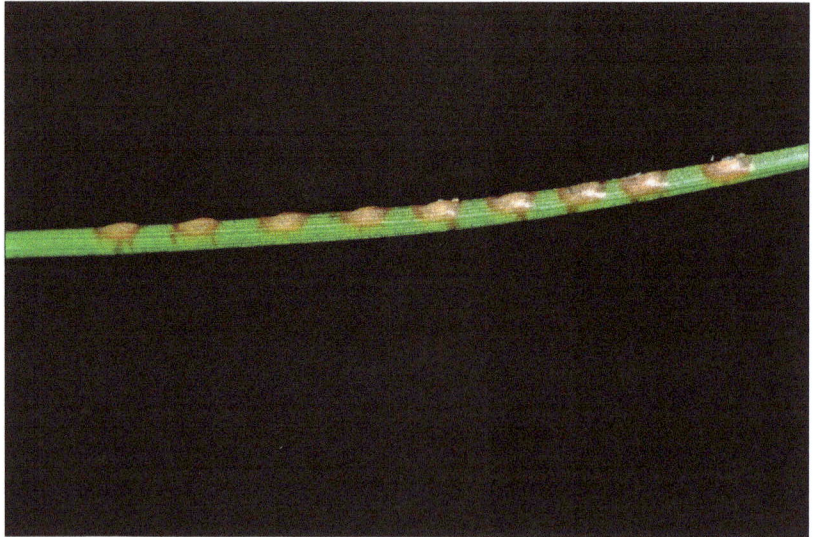

Egg slits in needle. (A.S. Munson, USFS, Bugwood.org)

Grasshoppers

Melanoplus spp.

Hosts: All Christmas tree species

Importance: Many species of grasshopper will eat Christmas tree foliage when field crops or other preferred vegetation is in short supply. Large numbers of grasshoppers can kill seedlings planted in grassy areas or devour the needles and bark of larger trees, making them unsalable for several years.

Look For:
- *Ragged needles* that have been partly or completely chewed off.

- *Scarred bark* on twigs and branches covered with hardened globs of pitch. Seedlings may be almost completely eaten.

MID-JULY TO OCTOBER
- *Large numbers of grasshoppers,* up to 1¼ inches long, feeding or resting on needles.

Biology: Grasshoppers become Christmas tree pests when their own food supply—grasses and field crops—is scarce. This happens most often during a drought.

Monitoring and Control: In areas where grasshopper injury is likely, examine trees of all ages regularly from mid-July through October. Treat if feeding damage is beginning to degrade trees. If noticed early enough, only the rows nearest the edge of a plantation may need treatment.

- Apply a registered insecticide directly to trees when grasshoppers are present, usually in August or September.

Next Crop:
- Reduce grassy vegetation on or near the site before planting.

Ragged needles and grasshopper. (USFS - NCRS Archive)

Gypsy Moth

Lymantria dispar

Hosts: Many hardwood species; spruces; occasionally pines, fir, and eastern redcedar

Importance: Gypsy moth larvae feed on the foliage of many trees, but rarely cause significant damage to Christmas trees. However, this exotic insect is regulated by Federal and State quarantines. Trees grown in States or counties known to be infested by gypsy moth must be inspected by regulatory officials before harvest. If gypsy moth egg masses or any other life stages are found during inspections, the field will be restricted and trees cannot be shipped outside of the infested area. In some infested States, trees that will be shipped to uninfested areas must be sprayed with approved insecticides at specific times to reduce the chance of introducing gypsy moth to new areas.

Look For:

ALL YEAR

- *Egg masses on the stem and branches of Christmas trees,* even when defoliation is not apparent. Gypsy moth egg masses are tan, covered with fine hairs, and may be 1 to 3 inches long. Egg masses are often on the underside of lower canopy branches.

MAY TO EARLY JULY

- *Gypsy moth caterpillars,* up to 3 inches long, with long hairs, blue and red spots on their backs, and mottled yellow and black heads.

Adult female and a fresh egg mass. (S. Katovich, USFS, Bugwood.org)

JULY TO AUGUST

- *Reddish-brown pupal cases,* often found in protected places on the stem or lower side of branches, in bark crevices, and on stems.

JULY TO SEPTEMBER

- *Adult moths.* Males are tan with darker markings, have feathery antennae, and are good fliers; females have white wings with black markings and rarely fly.

Biology: After hatching from eggs in early spring, tiny caterpillars climb to the tops of trees, drop off on a silken thread, and are blown about by wind currents. When they land on a suitable tree species, they

Caterpillar and pupa. (USFS Archive, Bugwood.org)

Gypsy moth caterpillar. (S. Katovich, USFS, Bugwood.org)

begin feeding. They prefer to feed on oaks, aspen, and other hardwood trees, especially when they are young. However, caterpillars will also feed on white pine, spruces, and other conifers. Feeding continues for about 6 weeks. Older larvae often wander about, searching for a dark, protected site for pupation, such as the stem of a Christmas tree. Adult moths emerge from pupal cases in 1 to 2 weeks. Female moths rarely fly and usually lay their egg masses near where they pupated.

Monitoring and Control: Gypsy moths are especially difficult to manage when plantations are near or adjacent to woodlots with oaks or aspen. Look for egg masses on trees in the woodlot and on Christmas trees bordering woodlots throughout the year. Watch for egg masses on the stems, undersides of branches, and in other protected areas.

If you have a gypsy moth population in your area, monitor egg masses in April and May to determine when egg hatch occurs. This will help you time your scouting and insecticide applications.

- See table 1 (page 22) for degree day information.

- Contact your State regulatory agency for current gypsy moth regulations in your area. In some States, trees must be sprayed with approved insecticides during the larval feeding period if the trees will be shipped out of an infested area.

- Look for egg masses. Be sure to look at the stem and turn over harvested trees to check for egg masses on the undersides of branches. Be especially vigilant when Christmas trees are grown near oak or aspen.

- Scrape off and destroy egg masses.

- When larvae are present, consider spraying trees with a registered insecticide.

- Monitor gypsy moth populations in woodlots adjacent to Christmas tree fields. Consider treating the borders of the woodlots with a registered insecticide if gypsy moth populations are high. Contact your county Extension office or regulatory agency for management options.

Next Crop:

- Plant new fields away from oak, aspen, and other stands dominated by species preferred by gypsy moth.

Introduced Pine Sawfly

Diprion similis

Hosts: Eastern white pine; less frequently Scotch, red, and Austrian pine

Importance: Larvae eat needles. Slightly defoliated trees might be degraded for a year or two, but severely defoliated ones are usually unfit for sale. This insect is usually not abundant enough to cause injury because natural enemies and low winter temperatures keep it at tolerable levels.

Look For:

- *Scattered patches of sparse or missing foliage* anywhere on the tree. The entire tree may be defoliated.

JUNE TO SEPTEMBER

- *Black-headed larvae,* up to 1 inch long, with yellow and white spots on a black background. They feed alone or with a few other larvae in a loose cluster on the needles.

- *Brown cocoons,* about ¼ inch long, among the needles at the base of small branches or on the tree trunk.

Biology: There are two generations a year. First-generation larvae begin feeding from late spring through early summer; they eat old needles from previous years. Second-generation larvae feed from late summer into the fall months, eating needles from the current year as well as older needles. The second generation is often larger and more damaging. Young larvae feed in small groups; older larvae disperse and

Empty cocoon attached to a twig. (S. Katovich, USFS, Bugwood.org)

Introduced pine sawfly larva. (S. Katovich, USFS, Bugwood.org)

are not clustered like most sawfly species.

Monitoring and Control: Inspect white pines of all ages from June to September. Look for larvae on trees scattered throughout the plantation. Treat young plantations (1 to 4 years old) when you notice 10 or more larvae per tree. Treat older plantations when larvae can be easily located and foliage injury is likely.

- See table 1 (page 22) for degree day information.
- Spray infested trees with a registered insecticide to control larvae.
- Spray foliage when larvae are small before damage is extensive, and spray only if larvae are present.

Next Crop:

- Avoid planting eastern white pine near windbreaks or stands of eastern white or Scotch pine.

Jack Pine Budworm

Choristoneura pinus pinus

Hosts: Scotch, red, jack, and Austrian pine

Importance: Budworm caterpillars eat needles and make them unfit for sale as Christmas trees. Lightly defoliated trees recover after a few years, but severely defoliated ones are degraded, attacked by other pests, or killed. Injury is most severe on trees that are beneath or near large, infested host trees, generally jack pine. This insect is usually important only within the range of jack pine in the Lake States.

Look For:

JUNE TO NOVEMBER

• *Defoliated shoot tips or branches with webbed clusters of brownish needles* attached to the twigs with silk. Most of the webbed needles wash off the tree by winter. Do not confuse with webworm feeding.

MID-MAY TO EARLY JULY

• *Caterpillars*, up to 1 inch long, feeding in the webbed foliage. Each has a black head and a characteristic black plate on the body segment behind the head. When mature, the body is reddish brown with small, cream-colored spots along the sides.

MID-JULY TO MID-AUGUST

• *Tan or brown pupae* or pupal skins within the webbed foliage.
• *Green egg masses* on the undersides of needles.

Biology: Jack pine budworms overwinter on trees as very small caterpillars. In early spring,

Jack pine budworm caterpillar. (S. Katovich, USFS, Bugwood.org)

they crawl out to the ends of shoots or are windblown to new hosts. They feed on current-year needles and opening buds, and attach uneaten, clipped portions of needles to the shoots with silk. When concealed in these clusters of webbed needles, budworms are difficult to control. Jack pine budworms have large cyclic outbreaks, so look for damage when you notice defoliated jack pine stands locally.

Monitoring and Control: If your Christmas tree field is adjacent to jack pine, inspect trees of all ages in mid-May when shoots first begin to expand. Consider treating the entire plantation if you find an average of 1 to 2 budworms per 10 shoots.

• See table 1 (page 22) for degree day information.
• Spray trees with a registered insecticide after larvae emerge in May or June.

Next Crop:

• Plant susceptible trees at least 500 feet away from jack pine windbreaks or stands.

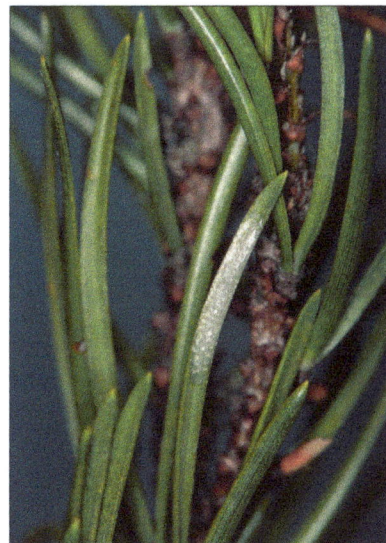

Hatched egg masses appear white. (USFS - NA Archive, Bugwood.org)

Young caterpillar on a pollen cone. (D.M. Benjamin, Univ. of WI, Bugwood.org)

Pine Chafer (Anomala Beetle)

Anomala oblivia

Hosts: All pines

Importance: Adult pine chafers damage new needles by gnawing through needle sheaths. Old needles and some bark tissues can be eaten after most of the new needles are destroyed. Affected needles bend or collapse, turn brown, and eventually drop, leaving a ragged display of shoots that degrades trees.

Look For:

JUNE TO SEPTEMBER

- *Broken or bent, green or brown needles.* Severely injured trees may look scorched in July because of brown needles. After August, the needles are short and have ragged ends.

- *Robust beetles*, ⅓ inch long, feeding on the shoots from mid-June to late July. The female beetle is tawny or buff; the male is brown with a greenish-bronze head.

Pests that cause similar symptoms: Pine needle midge can cause bent needles. Adult beetles can be confused with Japanese beetles, though Japanese beetles do not generally feed on pines.

Biology: Female beetles lay eggs in the soil of grassy areas near trees. The larvae feed only on grass roots and organic matter and do not harm trees. Adults emerge from the soil and begin feeding on needles in June and July, depending on location.

Monitoring and Control: Examine trees of all ages regularly in June. Treat the entire plantation if needles on trees within 3 years of harvest are bent, broken, or discolored. On younger trees, scattered light damage can be tolerated; treat only when damage impacts more than 20 percent of new needles.

- See table 1 (page 22) for degree day information.

- Shear to remove some of the injured foliage; the tree should outgrow the rest of the injury in 2 to 3 years.

- If control is needed, spray the trees once with a registered insecticide in late June when most of the beetles are feeding on the trees, but before they cause much injury.

Next Crop:

- Reduce grass cover in surrounding areas before planting.

Broken, green or brown needles. (USFS - NCRS Archive)

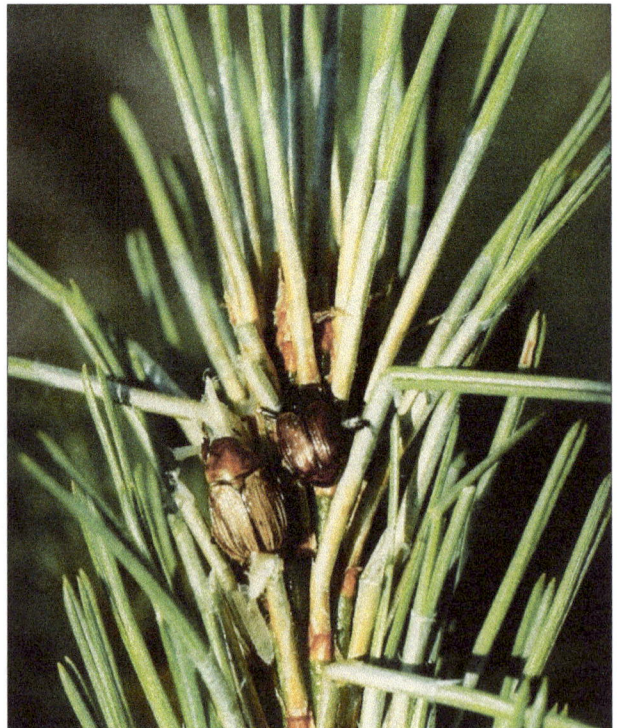

Female and male pine chafers on a shoot. (USFS - NCRS Archive)

Pine False Webworm

Acantholyda erythrocephala

Hosts: All pines; red and white pines are preferred

Importance: Pine false webworm is a sawfly that eats needles and forms conspicuous nests of clipped needles and insect frass held together with silk webbing. The larvae prefer to eat older needles, leaving the new year's foliage. This insect is not native to North America and occasionally damages pine plantations.

Look For:

- *Conspicuous, silken tubes with embedded, clipped needles and insect frass on branches and twigs.* Silk tubes are often wrapped around twigs and can be 4 to 6 inches long.

EARLY SPRING

- *Wasp-like adults* that are readily noticed flitting around from branch to branch, or crawling on needles. Adults are blue-black. Females have a bright orange head.

MAY TO JUNE

- *Pale-green larvae with purple stripes along the sides and back. Mature larvae have a yellowish head and are found inside their webbed nests.*

Pests that cause similar symptoms: Pine webworm is a caterpillar (not a sawfly) that has similar webbed nests.

Biology: The larvae construct loose, silken tubes along a twig from which they feed. They clip needles that they pull into their silk tubes. The tubes fill with needle parts and insect frass, creating the conspicuous webbed nests. Older larvae tend to feed as individuals, though several tubes can be found tied together. Larvae feed mostly on older foliage. There is one generation a year. Adults emerge in early spring, and eggs are laid on previous years' needles. Larvae feed through midsummer and then drop to the soil, where they remain until the following spring.

Monitoring and Control: Examine trees of all ages and treat when nests are too numerous to be removed by hand.

- Clip and/or destroy nests if they are few and scattered.

- Alternatively, spray trees with a registered insecticide when you see larvae feeding and building nests, generally in early to midspring. The insecticide *Bacillus thuringiensis* (Bt) is a control option for most caterpillars (moth species). Bt, however, does not kill sawfly larvae.

Pine false webworm larva. (B. Lyons, CFS, Bugwood.org)

Wasp-like adult and eggs. (B. Lyons, CFS, Bugwood.org)

Clipped needles tied with silk threads. (S. Katovich, USFS, Bugwood.org)

Pine Needle Midge

Contarinia baeri

Hosts: Scotch pine, occasionally red pine or other pines

Importance: The larvae of this European midge (small fly) feed at the base of needles, causing needles to droop, die, and drop prematurely. The bare leaders and thin crown caused by heavy feeding degrade the tree. Injured trees may be unsalable in the year of the attack, but will outgrow the injury in 2 to 3 years if the insect is managed.

Look For:

- *Needle loss*, mostly on the leader and shoots in the upper canopy.

MAY TO JULY

- *Needles bent over* that are either green or brown.

- *One or more small, yellow maggots*, 1/32 inch long, inside the papery sheath at the base of the needles (fascicle). Larval feeding causes a brown lesion between the needles inside the needle sheath. Remove the suspect needles and pull them apart to see the larvae and lesions. A hand lens may be needed.

Biology: Larvae overwinter in cocoons in the leaf litter and pupate in early spring. Adults emerge as the weather warms and fly to trees where they lay their eggs inside needle fascicles. Eggs hatch and larvae feed near the base of the needles, beneath the fascicle. Feeding causes the needles to bend over and eventually die. Full-grown larvae drop to the ground in midsummer and spin cocoons in the litter.

Pests that cause similar symptoms: Feeding by the pine chafer (Anomala beetle) can also cause needles to bend over and die.

Monitoring and Control: Verify that this insect is the cause of damage by locating the larvae or lesions between the needles. If more than 5 percent of the trees in a plantation show injury, consider treating the entire plantation the next spring as soon as larvae are found in the needle fascicles. Monitor for larvae from late May to late June, depending on latitude. For example, in southern Michigan, larvae can usually be found beginning in late May, while in northern Michigan larvae can be found starting around the third week of June. Control the insect as soon as you locate the first larva.

- See table 1 (page 22) for degree day information.

- Apply a registered insecticide when larvae begin feeding.

Yellow midge larva in a needle sheath. (USFS - NCRS Archive)

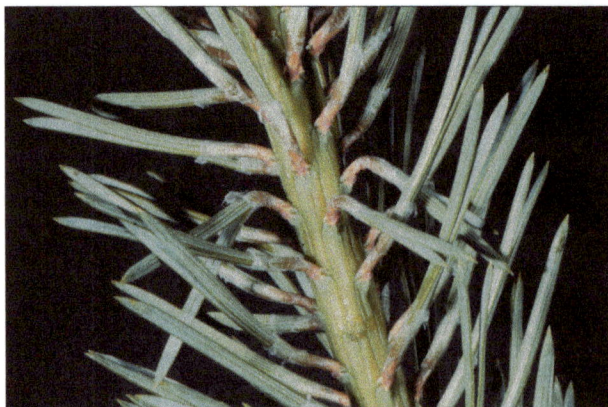

Midge feeding causes needles to bend or droop. (USFS - NCRS Archive)

Needle loss at the top of a tree. (USFS - NCRS Archive)

Pine Tube Moth

Argyrotaenia pinatubana

Host: Eastern white pine

Importance: Pine tube moth caterpillars bind needles together into a tube and feed on the needle tips. This injury is cosmetic and usually insignificant.

Look For:

- *Tubular clusters of 5 to 20 needles* bound with silk and squared off at the tips. Break the tubes open to find a yellow-green caterpillar or pupa, up to ½ inch long.

Biology: Pupae overwinter in the tubes and small moths emerge in early spring. Female moths disperse and lay eggs on needles, producing two generations during the summer. A single caterpillar lives within a tube of needles tied together with silk. They move to the end of the tube to feed on needle tips, slowly shortening the tube length to about 1 inch. Natural enemies can usually keep the population in check.

Monitoring and Control: Begin checking 2 years before harvest in the fall, winter, or early spring. Treat the entire plantation only if tubes become obvious enough to degrade trees.

- See table 1 (page 22) for degree day information.
- When practical, clip off and destroy tubes to kill the caterpillar or pupae.
- If it is necessary to control first-generation caterpillars, apply a registered insecticide between mid-May and mid-June while tubes are being formed (rarely necessary).
- Alternatively, apply a registered insecticide in mid- to late July to control second-generation caterpillars.

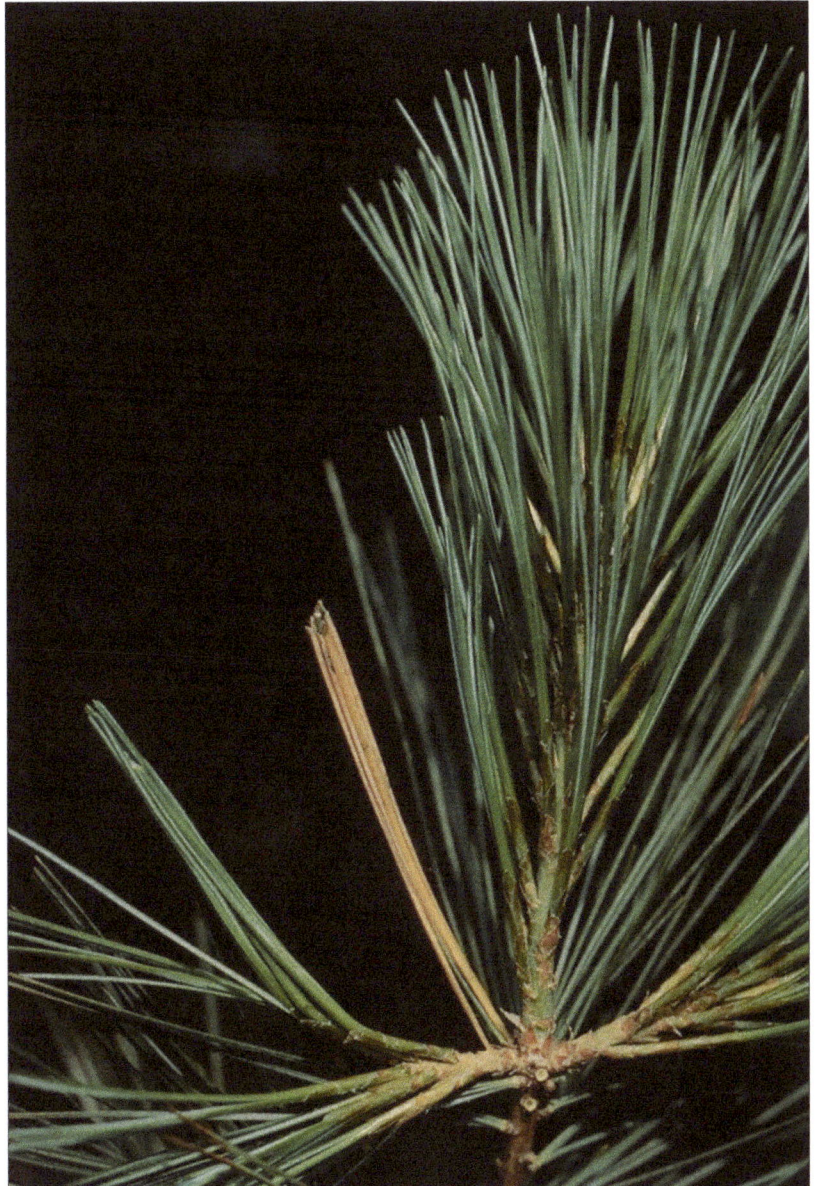

Tubular cluster of needles bound with silk and squared off at the end. (USFS - NCRS Archive)

Pine Tussock Moth

Dasychira pinicola

Northern Conifer Tussock Moth

Dasychira plagiata

Hosts: Eastern white and red pine, spruces, Fraser and balsam fir

Importance: These two related species look and behave very similarly. Outbreaks are not common but when they occur, they can be intense over a 1- to 2-year time period. Caterpillars can completely strip the needles from any size Christmas tree. Severely defoliated trees often die, and partially defoliated ones cannot be sold as Christmas trees. Although both tussock moth species occur throughout the Northeastern United States, tussock moth damage in general has only been a problem in localized areas in Wisconsin and Minnesota.

Look For:

- *Missing needles and ragged needle clumps* on some branches or on the entire tree. Check for pellets of insect waste on the ground beneath trees to be sure needles were eaten and have not merely fallen off. If they have fallen off, suspect a needlecast disease.

MAY TO EARLY JULY

- *Light-brown or reddish-brown caterpillars,* up to 1½ inches long, with four prominent tufts of gray hair on their backs.

JULY TO SEPTEMBER

- *Gray-brown, hairy cocoons or whitish egg masses*, about 1½ inches long, attached to needles.

Pests that cause similar symptoms: Redheaded pine sawfly can strip off both old and new needles in 1 year.

Biology: Caterpillars overwinter at the base of the needles or under the bark of the tree. In the spring they emerge and feed on needles. Feeding peaks in late June as the caterpillars mature. The insects then pupate, and adult moths soon emerge and lay eggs on the remaining needles. The eggs hatch in late summer, but the young caterpillars do not feed extensively until the following spring.

Monitoring and Control: Examine trees of all ages from May through early July, looking for feeding caterpillars and/or injured needles. Treat if caterpillars are abundant and easily located.

- Apply a registered insecticide to the trees as soon as caterpillars are noticed to avoid extensive feeding damage.

- Do not ship infested Christmas trees because overwintering caterpillars may emerge in the warmth of a home and feed on the tree during the holiday.

Next Crop:

- In the Lake States, avoid planting susceptible tree species near jack pine, a preferred host.

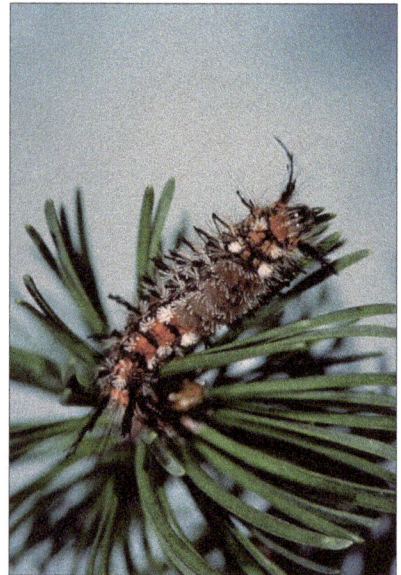

Pine tussock moth caterpillar. (USFS - NCRS Archive, Bugwood.org)

Northern conifer tussock moth caterpillar. (CT Agricultural Experiment Station Archive, Bugwood.org)

Pine Webworm

Pococera robustella

Hosts: All pines

Importance: Pine webworms are caterpillars that eat needles and sometimes form rather large nests of needle pieces and insect frass held together with silk webbing. The caterpillars prefer to eat older needles, leaving the new year's foliage.

Look For:

- *Conspicuous, loose nests of webbing, clipped needles, and insect frass* (brown pellets) that are 2 to 6 inches wide. Needles near the nest have been chewed off.

JULY TO AUGUST

- *Yellow-brown larvae*, ⅗ to ⅞ inch long, inside the nest. Many larvae can be in one nest. Larvae will have eight pairs of legs: three jointed pairs at the head end and five fleshy pairs at the back end.

Pests that cause similar symptoms: Pine false webworm is a sawfly that has similar webbed nests, although their nests tend to be smaller. The juniper webworm is another moth species that forms similar web nests on juniper, including redcedar.

Biology: There is generally one generation a year in the Northern United States. Pine webworm spends the winter in a cocoon in the soil. Moths appear throughout the summer and lay eggs on pine needles. The caterpillars occur in groups of up to 70 or 80 individuals. They form loose nests of insect waste (frass), clipped needles, and silk in which they live and feed. The caterpillars drop to the ground in late summer where they pupate.

Monitoring and Control: Examine trees of all ages and treat when nests are too numerous to be removed by hand.

- Clip and/or destroy nests if they are few and scattered.

- Alternatively, spray trees with a registered insecticide when you see larvae feeding and building nests.

- Large webbed nests can protect caterpillars from contact with insecticides, so spray for pine webworm as soon as caterpillars or nests are noticed.

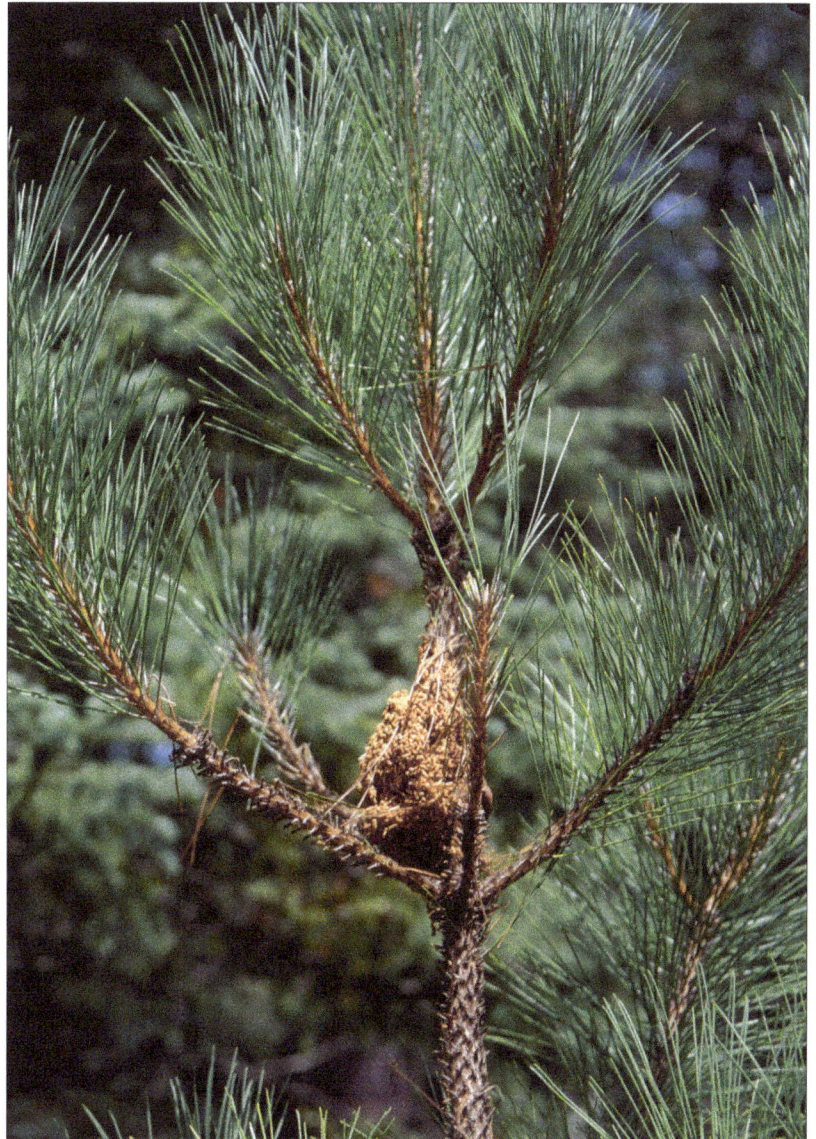

Loose nest of webbing, clipped needles, and fecal pellets. (S. Katovich, USFS, Bugwood.org)

Redheaded Pine Sawfly

Neodiprion lecontei

Hosts: Red and Scotch pine, occasionally spruce if it is interplanted with susceptible pines

Importance: Larval colonies strip the needles from Christmas trees, which kills branches, tree tops, or whole trees. Severely defoliated Christmas trees are unfit for sale. This insect becomes an important pest every 10 to 12 years and causes injury for 2 or 3 years before outbreaks subside. It prefers trees weakened by poor soil, drought, or competition from other plants.

Look For:

- *Sparse foliage on shoots or branches* anywhere on the tree. The entire tree may be defoliated.

JUNE AND AUGUST
(Central States)
OR JULY (Lake States)

- *Tufts of dry, straw-like needles.*

JUNE TO OCTOBER
(Central States)
OR JULY TO SEPTEMBER
(Lake States)

- *Yellow, black-spotted larvae* up to 1 inch long, with red heads, in clusters on the foliage.

Pests that cause similar symptoms: Pine tussock moth can strip trees of both new and old needles.

Biology: Redheaded pine sawflies in the Lake States usually have one generation whereas those in the Central States may have two or more generations. Pupae overwinter in cocoons spun in the litter or topsoil. Adults emerge in June in the north and in May and July farther south. Each female deposits 100 to 120 eggs in clusters on the needles. In 3 to 5 weeks, eggs hatch and

Straw-like needles. (USFS - NCRS Archive)

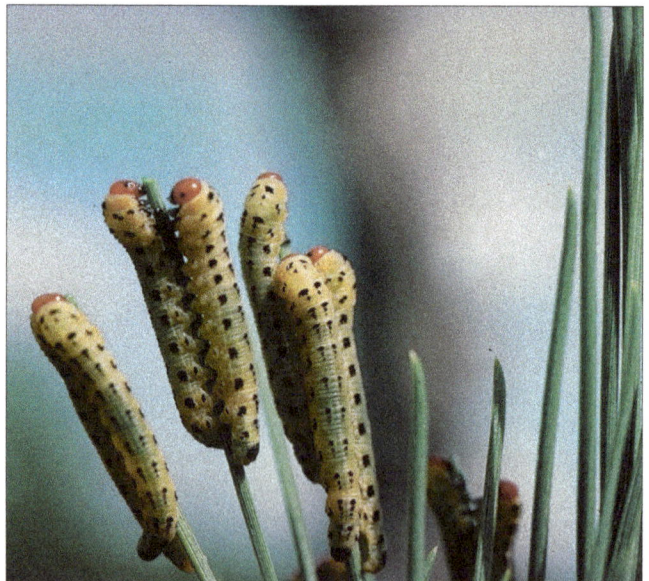

Redheaded pine sawfly larvae. (D. Mosher, MI DNR)

larvae begin to feed in groups. Larvae will eat both old and new needles, so trees can be completely defoliated in one season. One colony of 100 larvae can defoliate a tree 2 feet tall; 15 to 20 larval colonies can defoliate larger trees. Larvae feed for 5 or 6 weeks. After feeding, full-grown larvae drop to the soil and spin cocoons.

Monitoring and Control: Begin checking trees of all ages in June and continue through September (Lake States) or October (Central States). Treat individual trees as soon as you notice colonies. Treat the entire plantation when larvae are too abundant to control using hand methods.

- See table 1 (page 22) for degree day information.
- Knock occasional, scattered colonies of larvae off and crush them.
- Alternatively, when larvae are small, apply a horticultural oil, insecticidal soap, or registered insecticide that lists sawflies on the label.
- Control competitive plants, particularly bracken fern, with herbicides to increase tree vigor.

Next Crop:
- Destroy dense weeds and bracken fern with herbicides before replanting.
- Do not plant trees on dry, nutrient-poor soils.

Cluster of larvae. (L.L. Hyche, Auburn Univ., Bugwood.org)

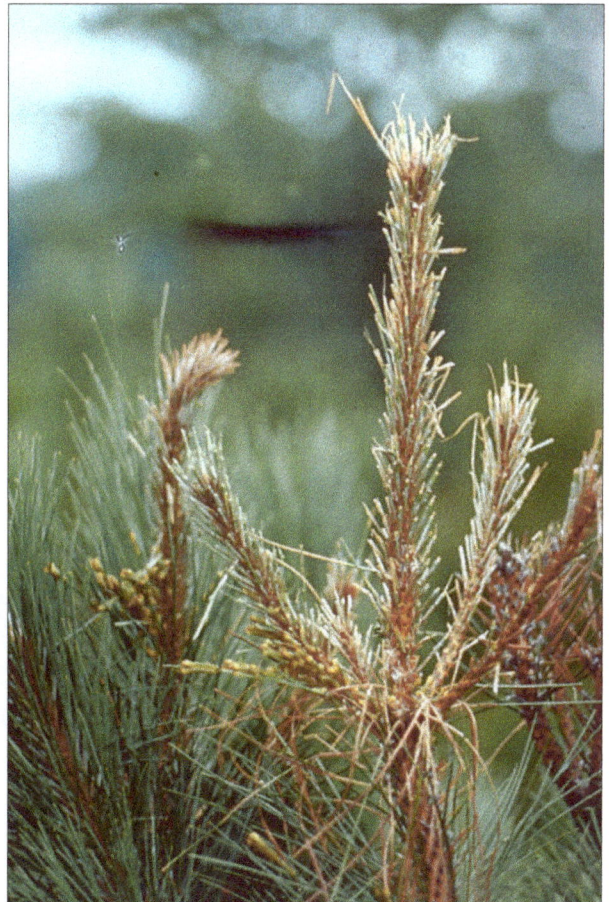

Feeding damage. (MN-DNR Archive, Bugwood.org)

Spruce Budworm

Choristoneura fumiferana

Hosts: All spruces and firs, occasionally pines growing intermixed with fir or spruce

Importance: Spruce budworm caterpillars defoliate trees, making them unfit for Christmas tree sale. Slightly defoliated trees recover after a few years, but severely defoliated ones are degraded, attacked by secondary pests, or killed. Injury is most severe on trees that are beneath or next to large, infested host trees. This insect is usually important only within the spruce-fir range in the Great Lake States and in the northern New England States where extensive natural stands of balsam fir and white spruce occur.

Look For:

JUNE TO NOVEMBER

- *Defoliated shoot tips or branches with webbed clusters of brownish needles* attached to the twigs with silk. Most of the webbed needles wash off the tree by winter.

MID-MAY TO EARLY JULY

- *Caterpillars*, up to 1 inch long, feeding in the webbed foliage. Each has a black head, characteristic black plate on the body segment behind the head, and a light-brown body when young. When mature, the body is gray brown with small, cream-colored spots along the sides.

MID-JULY TO MID-AUGUST

- *Green egg masses* on the undersides of needles.

Pests that cause similar symptoms: The yellow-headed spruce sawfly (not in this manual) can defoliate spruce trees. The balsam fir sawfly defoliates both fir and spruce. Neither of these web needles together or produce silk threads.

Biology: Spruce budworms overwinter as small caterpillars. In early spring, the young caterpillars become active and spin silk threads that get picked up in the wind, moving caterpillars to new host trees. Caterpillars begin feeding on needles and expanding buds, and attach the uneaten, clipped portions of needles to the shoots with silk webbing. When concealed in these clusters of webbed needles, budworms are difficult to control. Regional outbreaks in natural spruce-fir forests occur at irregular intervals, but when they do occur they last for many years and can be very destructive. Caterpillars only eat the new foliage in any one year, but persistent outbreaks can kill entire trees.

Monitoring and Control: Be aware of outbreak populations in any nearby spruce-fir stands. Inspect trees of all ages in May when buds first begin to expand. Treat the entire plantation when you find an average of 1 to 2 larvae per 10 spruce buds, or 1 to 2 larvae per 20 fir buds.

- See table 1 (page 22) for degree day information.
- Spray trees with a registered insecticide after larvae begin feeding in mid- to late May.

Next Crop:

- Plant trees at least 500 feet away from stands of mature balsam fir or spruce trees.

Spruce budworm caterpillar. (CT Agricultural Experiment Station, Bugwood.org)

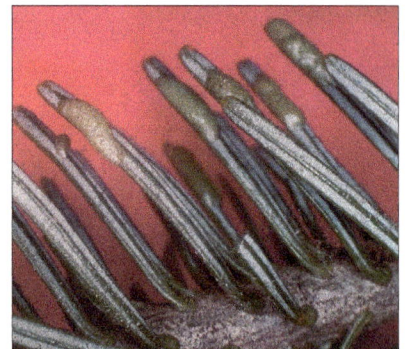

Egg masses on needles. (USFS - NCRS Archive)

Spruce Fir Looper

Macaria signaria

Hosts: Balsam and Fraser fir, Douglas-fir, eastern white pine, eastern hemlock, spruces

Importance: Caterpillars feed on needles, and heavy defoliation could affect the appearance of Christmas trees. However, this insect rarely causes severe defoliation. Occasionally, looper populations can build to damaging levels in early fall when growers may not be actively scouting fields.

Look For:

- *Brown needles, mainly in the upper crown*; they can, however, be found throughout the tree.
- *Portions of the needles that may be chewed off or notches in the needles that may be apparent.*

JUNE TO NOVEMBER

- *Looper or "inchworm" caterpillars on foliage*; caterpillars are light green with two light-colored, longitudinal stripes down the body and can be up to 1 inch long.

Biology: There may be one or two generations a year, depending on location and annual temperatures. Pupae overwinter in soil or debris below trees. Caterpillars can be present through the summer and into mid or late fall.

Monitoring and Control: Inspect plantations from summer through late fall. Look for both trees with brown needles and the distinctive inchworm caterpillars. Use a scouting board and rap shoots over the board. If caterpillars are present, frass (insect feces) will be obvious on the scouting board. Control is needed only if caterpillars are present in large numbers.

- Spray the foliage of infested trees with a registered insecticide when larvae (caterpillars) are present in high numbers.
- Spray foliage when larvae are small, before damage is extensive.

Needle feeding is concentrated on the top of the tree. (J. O'Donnell, MSU)

Spruce fir looper (caterpillar). (J. O'Donnell, MSU)

Closeup of needle feeding; note the notches. (J. O'Donnell, MSU)

Spruce Needleminers

Taniva albolineana, Epinotia nanana, Coleotechnites piceaella

Host: All spruce species

Importance: The caterpillars of several different needleminer species tunnel into the needles of spruce trees. Feeding by spruce needleminers will not kill a tree, but it can make Christmas trees look unsightly. Caterpillars cut and web needles into small, unsightly nests. Needleminers on spruce are common, but normally their populations are below a damaging level. Their activity rarely warrants treatment.

Look For:

- *Small clusters of discolored needles webbed tightly together* and flattened against the branch. Small trees may have damaged needles anywhere in the crown. The webbing on larger trees is found mostly on the inner parts of the lower branches.

- *Hollowed-out needles*, with a small hole usually near the base of each needle.

JUNE TO MID-APRIL

- *Brown, gray, or reddish larvae*, up to ¼ inch long, in the needles or within webbed foliage.

Biology: There are several different species of small moths that mine spruce needles; their life cycles tend to be similar. Caterpillars overwinter in hollowed-out needles or within clumps of clipped needles and waste held together with silk. In early spring they feed for a few weeks and then pupate. Adult moths begin emerging between mid-May (Central States) and mid-June (Lake States) and lay eggs on needles. In about 2 weeks, larvae hatch and begin mining needles. As they grow, they hollow out, cut, and web the needles together to form a nest-like enclosure.

Monitoring and Control: Inspect trees of all ages anytime during the growing season. Treat the entire plantation only if you notice injury on more than 10

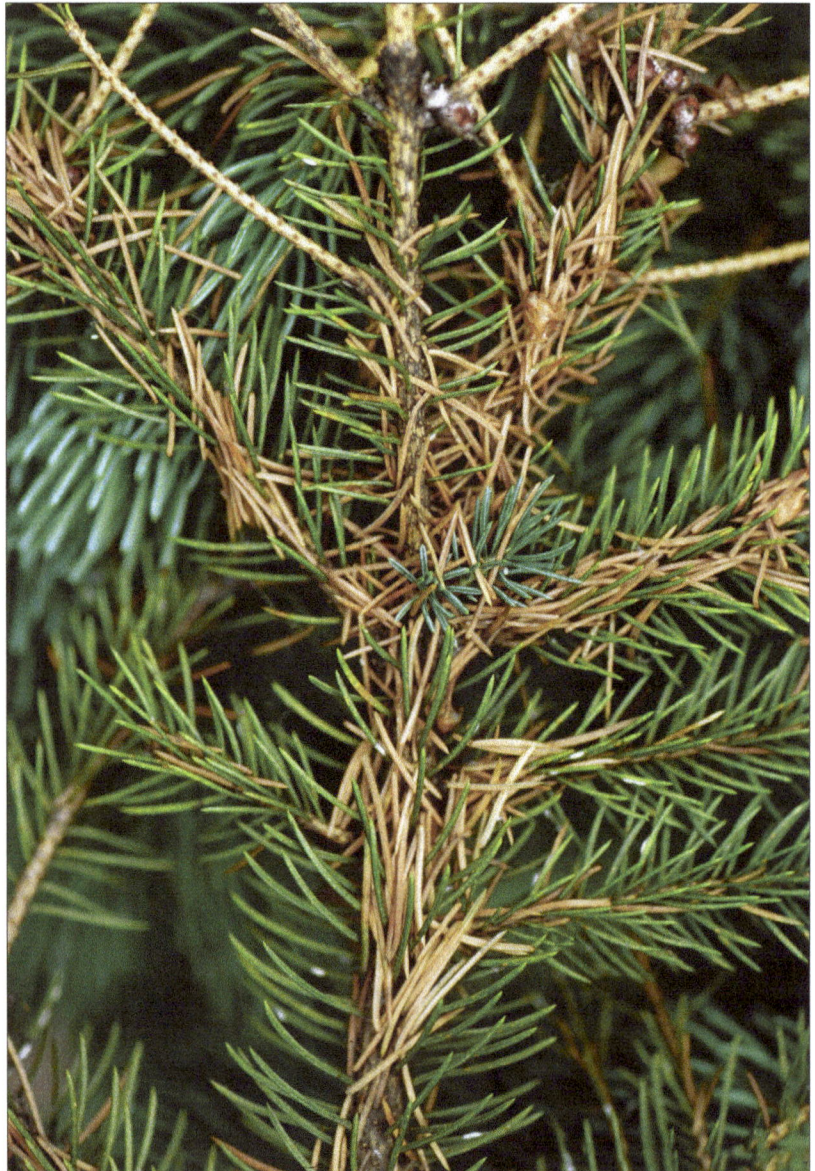

Clusters of discolored needles. (S. Katovich, USFS, Bugwood.org)

percent of trees that are within 3 years of harvest. On younger trees, treat only if needle damage is extensive and easily noticed.

- See table 1 (page 22) for degree day information.

- Spray trees thoroughly with a registered insecticide in mid-June (Central States) or mid-July (Lake States) to control larvae after they emerge from the eggs. Alternatively, spray trees in early spring after larvae begin feeding again.

Next Crop:

- Do not plant spruce near other heavily infested spruce.

Needles tied with webbing. (L. Gillman, USFS, Bugwood.org)

Spruce needleminers are small caterpillars. (CT Agricultural Experiment Station Archive, Bugwood.org)

Shoot/Branch Injury

Needles on shoots or branches uniformly discolored—usually red, yellow, or brown. Foliage may be black from sooty mold fungus. Frothy spittlemasses, aphid colonies, or scales may be on bark. If needles on shoots are cut off and webbed together, see previous section. If galls are present, see next section.

Aphids

Cinara spp., *Eulachnus agilis*

Balsam twig aphid is covered in the needle discoloration section.

Hosts: All Christmas tree species

Importance: Aphids suck sap from branches, shoots, and needles. Affected trees may become chlorotic (yellow), lose needles, attract secondary pests, and become unfit for sale. Aphids are very common on Christmas trees, though populations are most often below damaging levels.

Look For:

- *Discolored sparse foliage anywhere on the tree.* Scattered groups of needles turn yellow or red in summer and drop off in fall. Surrounding foliage may look sooty and glisten as if lacquered (black sooty mold). Bees and ants may be abundant on the foliage.

MAY TO NOVEMBER

- *Small, winged or wingless insects* clustered on the shoots or needles. Aphids may be yellow-green, brown, or black, and are usually about ⅛ inch long. The spotted pine aphid, which is greenish with black spots, grows to ¼ inch long. Some species are naked, and others are covered with a woolly wax.

Pests that cause similar symptoms: Trees infested with scale insects often develop black sooty mold. Adelgids are closely related to aphids and can be confused with them.

Biology: Most aphids overwinter on trees as eggs. Nymphs hatch in spring and quickly mature and reproduce. Several overlapping generations can produce large populations by late summer. Droughty weather at this time will increase needle fall. Predators such as lady beetles are important natural enemies of aphids and often keep aphid populations in check.

Monitoring and Control: Check trees during the growing season. Control may not be needed if lady beetles and other predators are abundant. Treat individual infested trees only if more than one-third of the shoots have aphid colonies. However, treat all infested trees if fields will be harvested in the current year.

- Limit the use of broad-spectrum insecticides that kill helpful aphid predators.

- Control mound ant colonies; these ants protect aphids by discouraging the natural enemies of aphids (see Allegheny mound ant).

- Spray trees if necessary with a registered insecticide to control large aphid populations. An insecticidal soap may be a good alternative to a conventional insecticide and will be less harmful to beneficial insects.

- Limit application of high-nitrogen fertilizers that can encourage aphid outbreaks.

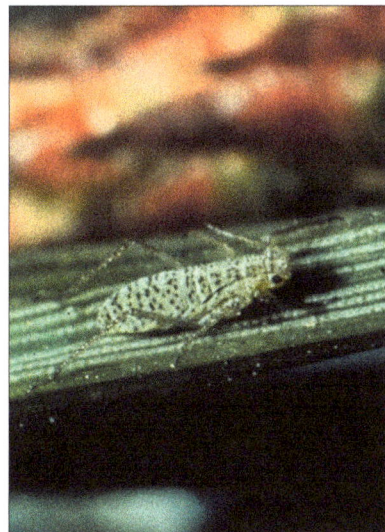

Spotted pine aphid on a needle. (D. Mosher, MI DNR)

White pine aphids on a stem. (D. Shetlar, Ohio State Univ.)

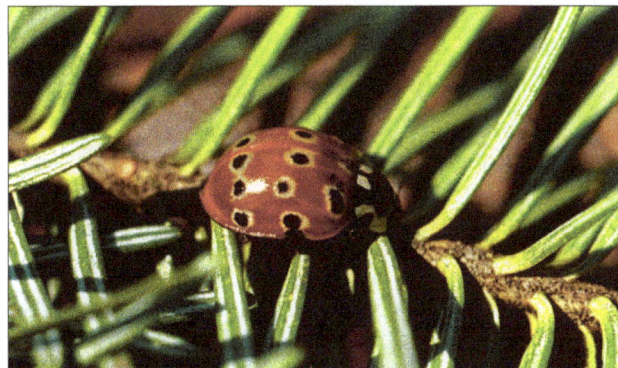

Lady beetle, a common aphid predator. (W. Cranshaw, Colorado State Univ., Bugwood.org)

Balsam Shootboring Sawfly

Pleroneura brunneicornis

Hosts: Balsam and Fraser firs

Importance: Heavily attacked branches develop many small buds behind the killed shoots, which can distort branch growth. Egg-laying adults prefer trees that break bud later in the spring, so Fraser fir is often more heavily attacked than earlier bud-breaking balsam. This sawfly is more common in the Northeastern United States than in the Western Great Lakes Region.

Look For:

- *Somewhat flattened, wilted new shoots.*
- *Needles on damaged shoots that turn from pale yellow to reddish brown.*
- *Shoots that pull off easily.*
- *A white larva inside a hollowed-out shoot axis.*
- *Clusters of small buds at the base of damaged shoots.*

Adult sawflies can be hard to observe in the field. They are most easily seen just before buds on fir begin to swell. Look for them flying around the crowns of trees that have been previously heavily attacked.

- *Adults look like flies, about 1/5 inch long, with distinctive black and white stripes on their abdomen.*

Pests that cause similar symptoms: Frost damage, though shoots damaged by the shootboring sawfly pull off easily, revealing a hollowed-out shoot.

Biology: Larvae overwinter deep in the soil and emerge as adults in early spring. Most larvae spend 2 years in the soil, so local populations are often high one year and low the next. Adults lay eggs beneath the bud sheath just before buds swell. The timing between bud development and egg laying is critical. Adults emerge well before any visible signs of fir growth but delay laying eggs until temperatures warm into the 60s (°F) and buds are at the right stage of development. Early bud swell, often by balsam fir, may occur before adults have begun to lay eggs. Later bud swell by Fraser fir tends to occur when conditions are more favorable for egg laying. Larvae tunnel into the center of the developing shoot, then cut an exit hole and drop to the ground by late spring.

Monitoring and Control: If adults are easily seen, expect noticeable damage. In late spring and early summer, look for flattened shoots that are reddish brown. These shoots break off, so surveys later in the year will miss this injury. Given that heavy damage is uncommon, especially for consecutive years, control measures are rarely warranted.

If damage is deemed excessive:

- To control adults, spray infested trees with a registered insecticide just before buds begin to swell or when eggs are being laid.
- To reduce future population levels, spray infested trees with a registered insecticide just prior to larvae dropping to the ground. This is about the time lilacs are in full bloom.

Characteristic flattened shoot on balsam fir. (R.S. Kelley, VT Dept. of Forests, Parks and Recreation)

Fly-like adult sawfly laying eggs. (R.S. Kelley, VT Dept. of Forests, Parks and Recreation, Bugwood.org)

Balsam Woolly Adelgid

Adelges piceae

Hosts: All species of fir; Fraser and balsam fir can be seriously affected

Importance: Where present this small insect can cause significant damage to fir Christmas trees. Balsam woolly adelgid is not native to North America. It is currently established in the Northeastern United States and adjacent Canada, the Pacific Northwest, and the Appalachian mountains, where it has devastated natural stands of Fraser fir. New introductions could occur most likely through the movement of infested fir nursery stock.

Look For:

These symptoms of advanced infestation are often obvious:

- *A flat top or weak terminal that is slanted.*
- *Swollen twigs that drop their needles* (referred to as gouting).
- *Dead shoots or branches.*
- *Wilted appearance of shoots.*

The adelgids themselves are small and difficult to see. Look for:

- *White wool* produced by adults as they feed, usually on the trunk below where branches emerge. Wool often remains on the bark throughout the year.

Pests that cause similar symptoms: Pine bark adelgid infests white pine and occasionally other pines. Balsam twig aphids may be found on fir needles; they produce some white, woolly wax.

Biology: Adelgids are sedentary for most of their life cycle. Eggs hatch into crawlers, the only mobile life stage. Crawlers can disperse short distances but may be blown or perhaps carried by birds to new host trees. The crawlers settle on the bark, insert their straw-like mouthparts, and begin feeding on sap, eventually maturing into adult females. No

Swollen twigs or stems. (R.S. Kelley, VT Dept. of Forests, Parks and Recreation, Bugwood.org)

Eggs wrapped in white wool. (R.S. Kelley, VT Dept. of Forests, Parks and Recreation, Bugwood.org)

males occur in North America. As the adelgids feed, they produce a woolly covering of wax. Mature females lay eggs under the woolly covering. There can be two or three generations per year depending on location and elevation. Adelgids overwinter as immature nymphs. During the summer, life stages can overlap.

Monitoring and Control: Watch for balsam woolly adelgid even in areas where it is not currently known to occur. Closely examine any fir tree that shows symptoms of infestation. Balsam woolly adelgid is not known to be established in States in the North Central Region. If you suspect a tree may be infested, notify State regulatory officials. The best control for this insect is to keep it out of your plantations.

- Buy stock only from reputable nurseries.
- Do not transport wild trees from infested areas.

If you find adelgids, treatment will be needed. Shipping infested trees to areas not yet infested can start new infestations and result in fines and penalties.

- Thoroughly spray infested trees with a registered insecticide. Complete coverage and good penetration of spray onto the bark surfaces are important.
- Cut and destroy any heavily infested trees. Be careful to avoid spreading the crawlers when moving infested trees to a burn pile.
- Eliminate large, infested fir trees in adjacent stands or infested fir growing in nearby abandoned Christmas tree fields.

Next Crop:
- If this is a serious problem in your area, consider growing nonhost species.

White, woolly flecks on bark. (S. Tunnock, USFS, Bugwood.org)

Broom Rust of Fir

Broom Rust of Fir

Melampsorella caryophyllacearum

Hosts: Balsam, Fraser, and white fir
Alternate Host: Chickweed

Importance: This disease causes the formation of witches' brooms (dense, bushy masses of host branches) within the trees, which reduces their quality. Because the incidence of infected trees is usually very low within a plantation, this disease is rarely considered a major threat to Christmas trees.

Look For:
JUNE
- *Short, thick, upright shoots with stunted, thickened, pale-green needles.*

JULY TO AUGUST
- *Yellow needles in the broom* with orange-yellow pustules on the underside of the needles.

SEPTEMBER
- *Needles in the broom falling to the ground*, leaving a mass of stunted, thick shoots.

Biology: Fir buds are infected in the spring by windblown spores from chickweed, and the fungus grows into the branch where it overwinters. One year after infection, stunted, thick shoots grow upright out of the infected branch. Spores are produced on the needles within the broom. Infected shoots within the broom will produce new needles each year, and the fungus will in turn produce spores on these needles in the summer. These spores infect chickweed, the alternate host, but not other fir trees. This disease is only a problem when chickweed is present.

Monitoring and Control:
Examine your trees in July and August. The yellow brooms will be conspicuous.

Disease incidence will rarely reach a level that causes concern. If it does, consider controlling chickweed within the plantation.

- Remove brooms from infected trees. Once the foliage and woody broom material dries out, the fungus will die.
- If disease incidence is high, mow or kill chickweed in and around the plantation.
- There are currently no fungicides labeled for control of broom rust.

Next Crop:
- Inspect nursery stock before planting; do not plant infected trees.
- Examine areas around potential plantation sites. If broom rust is present in nearby native balsam or Fraser fir, be prepared to accept some level of disease, or plant species other than balsam, Fraser, or white fir.

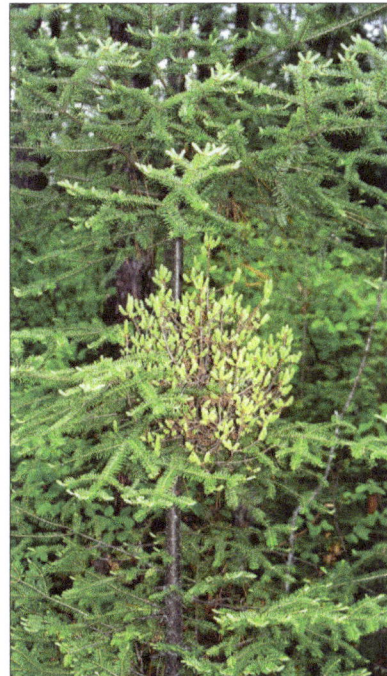

Witches' broom on fir caused by the rust fungus. (S. Katovich, USFS, Bugwood.org)

Rust pustules on infected needles. (J. O'Brien, USFS, Bugwood.org)

Deer

Odocoileus virginianus

Species Affected: All pines; Douglas-fir, Fraser fir, and other true firs; Norway spruce and white spruce

Importance: Deer feed on the shoots of young trees. In large numbers, these animals can cause extensive damage to tree plantings. Heavily browsed trees may be too deformed to be sold as Christmas trees.

Look For:

- *Ragged, squared-off ends of deer-browsed twigs*, ½ inch or less in diameter, on the lower 6 feet of the tree.

- *Deer droppings and tracks* near trees.

- *Strips of shredded or damaged bark on stems* of large trees. Bucks may rub the bark off when polishing their antlers against stems in early fall.

Pests that cause similar symptoms: Rabbit and hare will clip off buds and twigs; however, they clip twigs with a neat, clean cut.

Monitoring and Control: Examine trees of all ages throughout the year even though most damage occurs in late winter or early spring. Contact a conservation officer or a wildlife pest control specialist to help determine if reducing the deer population is appropriate under local conditions. No control is needed if injury is random and infrequent.

- Consider fencing fields with permanent fencing to exclude deer.

- Use a deer repellent for short-term control (2 to 6 weeks) when appropriate. Keep in mind that some deer repellents are often ineffective, and all must be reapplied frequently.

- In some States, wildlife management agencies may offer programs to help landowners reduce deer numbers. Contact the appropriate agency for information on crop depredation programs.

Next Crop:

- Avoid planting vulnerable species where deer are known to congregate during the winter. Consider planting Colorado blue spruce or balsam fir in vulnerable areas.

- Consider fencing fields to exclude deer if damage is often severe.

From left to right: Seedling browsed by deer. (MN-DNR Archive, Bugwood. org)

Damage caused by deer polishing their antlers. (J. Liska, Forestry and Game Management Research Institute, Bugwood.org)

Delphinella Shoot Blight

Delphinella Shoot Blight

Delphinella balsameae

Hosts: Balsam, white, grand, Fraser, subalpine, and Noble firs

Importance: This fungus kills young shoots. Disease is more severe in the lower crown. Severely affected trees can become stunted or killed after repeated years of infection.

Look For:

SRING and SUMMER

- *Pinkish spots on new needles. Infected needles soon turn red.*

- *Shriveled needles* that give the appearance of being only half the normal width.

- *Dead, young shoots*, especially lateral shoots. Needles become brittle but are retained for up to 2 years.

- In midsummer, *black fruitbodies* that erupt through upper needle surfaces of killed shoots.

FALL and WINTER

- Shriveled, dead needles attached to dead shoots.

Pests that cause similar symptoms: Rhizosphaera needle blight of firs, frost

Biology: Spores produced in fruitbodies on attached and fallen infected needles are dispersed in spring, infecting elongating shoots. Symptoms develop in the spring and summer, and the fungus overwinters in the diseased needles, completing the 1-year disease cycle.

Monitoring and Control: Check trees of all ages throughout the year.

- If you are at risk of significant economic losses, consider spraying with a protectant fungicide at bud break and again 10 to 12 days later.

- Remove severely affected shoots; dead seedlings; and any diseased, wild border trees to reduce sources of the fungus.

Next Crop:

- Plant only disease-free nursery stock.

- Inspect trees carefully for the first few years after planting for disease symptoms.

Red, infected needles. (R.S. Kelley, VT Dept. of Forests, Parks and Recreation, Bugwood.org)

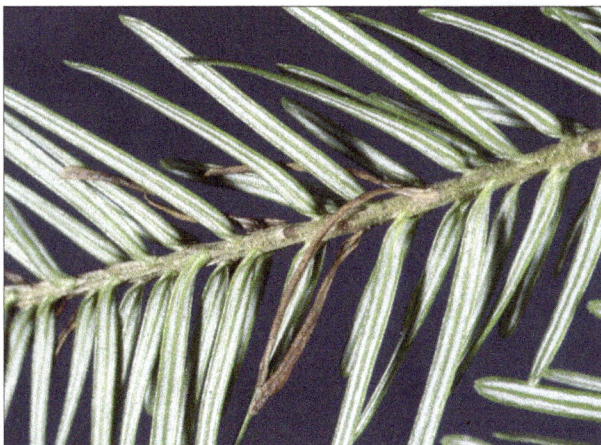

Shriveled, infected needles. (R.S. Kelley, VT Dept. of Forests, Parks and Recreation, Bugwood.org)

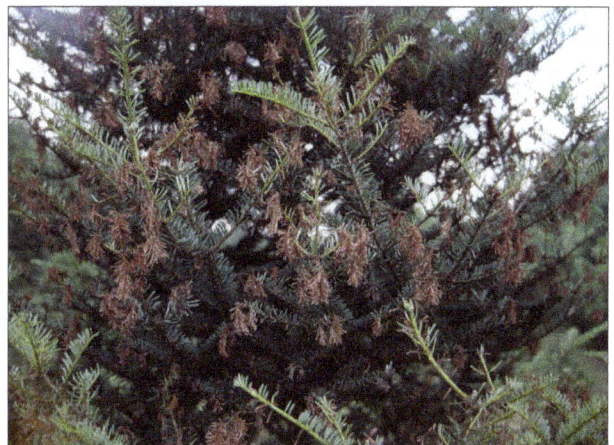

Infected branches with red needles. (R.S. Kelley, VT Dept. of Forests, Parks and Recreation, Bugwood.org)

Diplodia (Sphaeropsis) Shoot Blight and Canker

Diplodia pinea

Syn. *Sphaeropsis sapinea*

Hosts: Red, Scotch, and Austrian pine

Importance: This fungus kills current-year shoots on trees of all ages and usually kills nursery seedlings within the first year. Repeated infection over several years eventually kills older trees. The fungus can also cause girdling cankers on stems and branches of affected trees. *Diplodia* is an opportunistic fungus that can severely damage drought-stressed trees or trees wounded by insect pests, hail, or other agents.

Look For:

- *Stunted or curled current-year shoots.* Infected tissue will be soaked with resin.
- *Black fruitbodies* on dead needles and shoots.
- *Cankers (resinous sunken areas)* on branches or stems. The portion of the tree above girdling stem cankers will be killed.
- *Olive-green streaking* on the resin-soaked tissue beneath the bark of affected branches and stems.

Pests that cause similar symptoms: Drought, European pine shoot moth, Nantucket pine tip moth, pales weevil, pine root tip weevil, pine shoot beetle, Scleroderris canker, spittlebugs, Sirococcus shoot blight

Biology: *Diplodia* overwinters on pine in needles, shoots, bark, and cones, and in pine litter on the ground. The fungus infects elongating shoots in the spring. Spores are released during wet weather throughout the growing season. Wounds, such as those made by hail, shearing, or insects such as the pine spittlebug, serve as entry points for *Diplodia*. The fungus, however, does not need wounds to infect trees. Once it has infected its host, it can persist without causing symptoms. Disease and mortality are triggered by a stress event, such as a drought, hail, or outplanting stressed seedlings.

Monitoring and Control: Inspect trees of all ages in late spring or early summer.

Tree damage caused by Diplodia shoot blight. (L. Haugen, USFS, Bugwood.org)

Randomly select trees and look for stunted or curled current-year shoots. If trees are unsalable because of damage, consider treating the entire plantation the next spring. Take other preventive measures immediately to avoid spreading the fungus.

- Provide water during drought.

- Control insect pests that are weakening trees and creating entry points for *Diplodia*.

- Do not shear infected trees during wet weather because spores released at this time may be carried from tree to tree on shearing tools.

- Prune out infected branches and disinfect tools between cuts.

- Apply a registered, preventive fungicide four times, once every 2 weeks during shoot elongation, to prevent *Diplodia* from spreading to healthy trees. Make the first application just prior to bud break.

Next Crop:
- Plant disease-free stock. *Diplodia* can persist in asymptomatic seedlings so it is important to be sure the seedlings have been protected from infection in the nursery.

- Avoid planting susceptible species, such as Austrian or red pine, on poor sites where they can become stressed.

- Do not plant trees next to windbreaks that are affected by *Diplodia*. Examine windbreaks closely; although shoots may not be infected, cones may still harbor *Diplodia*.

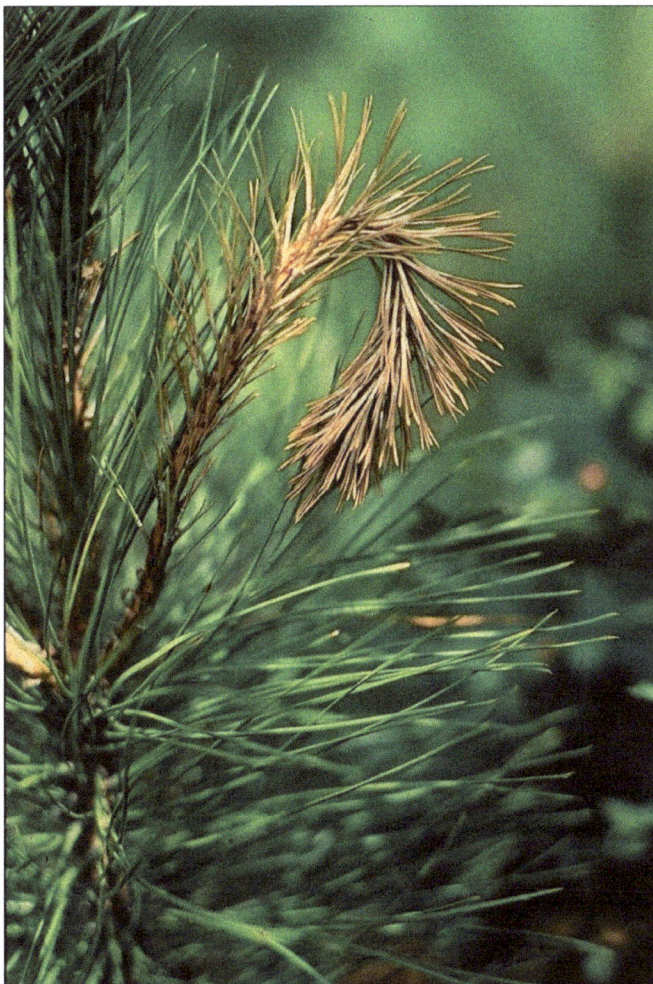

Diplodia-infected current-year shoot. (USFS - NCRS Archive)

Black fruitbodies of Diplodia on dead needles. (USFS - NCRS Archive)

Eastern Pine Shoot Borer

Eucosma gloriola

Hosts: Eastern white and Scotch pines are preferred; other pines, white spruce, and Douglas-fir can be hosts

Importance: The caterpillar of this small moth usually attacks new lateral (side) shoots. When abundant, larvae can damage the general shape of the crown by killing many shoots. The terminal leader is sometimes attacked, which affects tree form.

Look For:

JUNE TO OCTOBER

- *Flagged (discolored) shoots* on pine and spruce. The 6- to 8-inch-long ends of shoots turn yellow and then red. Douglas-fir shoots wilt and droop before yellowing, curling into the shape of a shepherd's crook.

- *Terminal leaders or branch ends broken over near their bases* that leave distinctive, flat stubs.

- *An oval hole at the base of the injury* through which the caterpillar has escaped.

- *Caterpillars feeding down the pith or center of shoots, packing sawdust-like frass into the tunnel.* Find caterpillars by cutting an infested shoot lengthwise with a knife. If cut before mid-July, you may find a single, dirty white to gray larva, up to ¾ inch long, in the shoot.

Pests that cause similar symptoms: Diplodia shoot blight, European pine shoot moth, jack pine tip beetle, and pine shoot beetle kill shoots. White pine weevil only infests the terminal shoot.

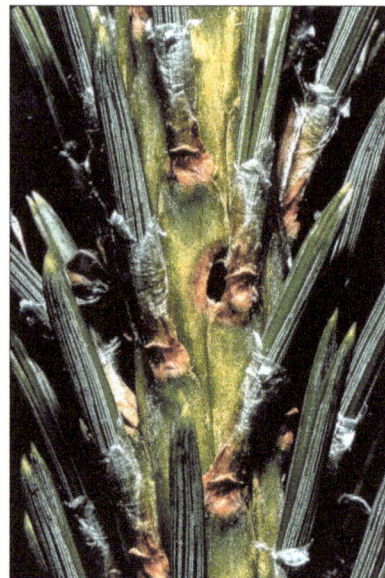

Oval exit hole in a damaged shoot. (USFS - NCRS Archive)

Flagged, crooked shoots. (USFS - NCRS Archive)

Biology: Female moths emerge in May and lay eggs on the new shoots. Most attacks occur on shoots in the upper crown. Young larvae bore into and feed in the center of elongating shoots. After chewing an oval hole at the base of the feeding tunnel, mature larvae emerge and drop to the ground to pupate and overwinter.

Monitoring and Control:
Examine trees of all ages from midsummer to frost. If you find more than 10 injured shoots per tree and trees are within 2 years of harvest, plan to treat the entire plantation the next spring.

- See table 1 (page 22) for degree day information.

- Apply a registered insecticide in mid-May to kill larvae before they bore into shoots.

- Do not delay treatment in early spring. By the time injury is apparent, most larvae have left the shoots, and control will not be effective.

- Early shearing can kill larvae inside shoots. Correctively prune forked branches by removing excess terminals. Clip off damaged shoots on trees before marketing.

Next Crop:
- Plant resistant varieties of Scotch pine, such as Swedish, Riga, or Scandinavian.

Frass packed into a feeding tunnel. (USFS - NCRS Archive)

Tunneled shoots die and break off. (USFS - NCRS Archive)

Eastern Pine Weevil

Pissodes nemorensis

Hosts: All pines and spruces

Importance: This insect causes damage where weak or dead pines are left standing, or where many fresh stumps are available for weevil breeding. There are two distinct types of damage: adult weevils chew holes in the bark, and larvae can infest young trees where they feed just under the bark in the main stem. Feeding by adults may kill some shoots, leaving dead spots in the crown and slightly degrading Christmas trees. Larvae may be found killing young trees (2 to 5 years old) that are growing slowly, generally because of heavy grass competition.

Look For:
- *Flagged (discolored and deformed) shoot tips* anywhere on trees or seedlings.
- *Small, circular feeding wounds ("drill holes")* at the base of injured shoots. Pitch may ooze from the wounds.
- *Small, white pupae or C-shaped larvae*, ⅓ inch long, beneath the bark of dead trees or stumps.
- *Elliptical chambers in the wood* beneath the bark, each covered with a ½-inch-long chip cocoon made of fine wood shavings.
- *Light-brown, white-spotted weevils*, ⅓ inch long, feeding on pine shoots after dark.

Pests that cause similar symptoms: Adult Pales weevil feeding causes more widespread damage by stripping off patches of bark. Small trees can be killed by a variety of agents including voles, pocket gophers, Armillaria root disease, blister rust, drought, and Diplodia canker.

Biology: Female weevils lay eggs in the spring on the inner bark of stumps and recently dead, dying, or severely stressed trees. Larvae feed under the bark and make chip cocoons in which to pupate. After pupating in late summer, adult weevils emerge and feed on the inner bark of twigs and small branches. They overwinter in the litter around infested trees.

Monitoring and Control: Inspect trees of all ages, especially where fresh stumps are available. In early summer, look for pitchy "drill holes" or flagged tips on seedlings or older trees. Treat the entire plantation if feeding damage on seedlings is readily observed or if older trees have five or more flagged tips per tree.

Eastern pine weevil adult. (USFS - NCRS Archive)

Distinctive "chip cocoons" found under the bark. (C. Evans, IL Wildlife Action Plan, Bugwood.org)

Infestations lighter than this do not need control because most flagged shoots will fall off before harvest.

- See table 1 (page 22) for degree day information.
- Remove, chip, or burn dead or dying pines and fresh stumps before late spring to eliminate breeding material.
- Alternatively, in April to mid-May, drench fresh stumps and nearby soil once with a registered insecticide to kill the egg-laying adults. A similar drench in August will kill emerging adults.
- Spray living trees once with a registered insecticide in August or September to kill the feeding adults.
- When harvesting, leave one whorl of live, pest-free branches on each stump to keep it alive and therefore unattractive to the weevils. Destroy these stumps within 3 years. Caution: Do not leave live branches that have needlecast diseases.
- Control grass competition around young trees.

Next Crop:
- Delay replanting a cutover area for 2 years unless stumps are removed or treated to prevent weevil attack.

Adult feeding injury on twigs forms dried beads of pitch. (C. Sadof, Purdue Univ.)

European Pine Shoot Moth

European Pine Shoot Moth

Rhyacionia buoliana

Hosts: Scotch, red, and Austrian pine

Importance: The caterpillars of this moth bore into the buds and developing shoots of host pines. This kills or deforms shoots, stunts growth, and can make Christmas trees unfit for sale. This insect's range is limited by cold weather. It is rarely found in the northern half of Wisconsin and Minnesota, or in the western half of Upper Michigan.

Look For:

- *Dead, stunted, or stubby shoots* anywhere on the tree. Shoots usually die before needles expand.
- *Hardened globs of pitch* where larvae have bored into shoots.
- *Trees that are distorted, bushy, or have multiple leaders.*

MID-APRIL TO EARLY JUNE

- *Brownish caterpillars with black heads,* up to ⅝ inch long, on or inside new shoots.

Pests that cause similar symptoms: Diplodia shoot blight, Nantucket pine tip moth, and pine shoot beetle can cause dead shoots and damaged buds.

Biology: In the spring, caterpillars bore into newly developing shoots to feed and pupate. This distorts new shoot growth. Moths emerging in June and July lay eggs that soon hatch into small caterpillars that bore into needles and buds to overwinter. During particularly cold winters, only caterpillars that are insulated on branches below the snowline will survive. Dry weather and poor soil conditions encourage population buildup.

Monitoring and Control: Inspect trees of all ages in April and May. Randomly select 30 to 50 trees scattered throughout a plantation and look for caterpillars and damage. Consider treating the entire plantation if 5 percent or more of the trees are injured. See table 1 (page 22) for degree day information.

- If attacks are light and scattered, prune and destroy attacked shoots before June.
- Wait to shear trees until mid-July when shearing will remove most of the eggs or larvae on shoot tips.
- Prune crooked, injured leaders and branches while shearing to restore trees to good form.
- Remove the lower whorl(s) of branches to prevent larvae from overwintering below the snowline and surviving. This will also give the trees "handles" and make harvesting easier.
- If necessary, spray trees with a registered insecticide during very early spring to kill larvae as they migrate to new shoots. Trees can be treated again in late June or early July after larvae hatch from eggs.

European pine shoot moth. (USFS - NCRS Archive)

Distorted growth. (A.S. Munson, USFS, Bugwood.org)

Bud damaged by caterpillar feeding. (M. Zubrik, Forest Research Institute - Slovakia, Bugwood.org)

Frost Injury

Species Affected: Balsam fir, Fraser fir, Douglas-fir, spruce; occasionally pines

Importance: Below-freezing temperatures in early spring can kill emerging shoots and degrade Christmas trees. Susceptible trees may become stunted or bushy if injured by frost several years in a row. Balsam fir can be very susceptible to damage.

Look For:

MAY TO JUNE

- *Brown, wilting, and dying shoots* of the current-year growth. New shoots will develop next to the dead ones.

- *Live, crooked shoots.*

AUGUST TO OCTOBER

- *Dead shoots remaining on trees* until late autumn. They may not drop until spring.

Monitoring and Control: Examine trees of all ages after a late spring frost. Take recommended actions if any noticeable damage occurs.

- Remove dead shoots when shearing.

- Harvest trees growing in frost pockets (low areas where cold air collects) as soon as possible.

Next Crop:

- Avoid planting highly susceptible species, such as true firs and Douglas-fir, in frost pockets. Pines may prove to be less susceptible. However, most frost pockets are not good sites for growing any conifer, so avoid them.

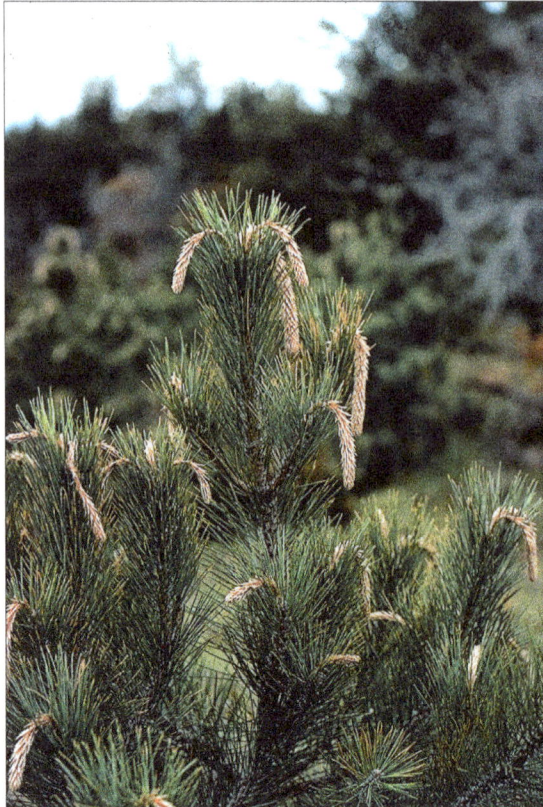

Frost injury to developing shoots. (USFS - NCRS Archive, Bugwood.org)

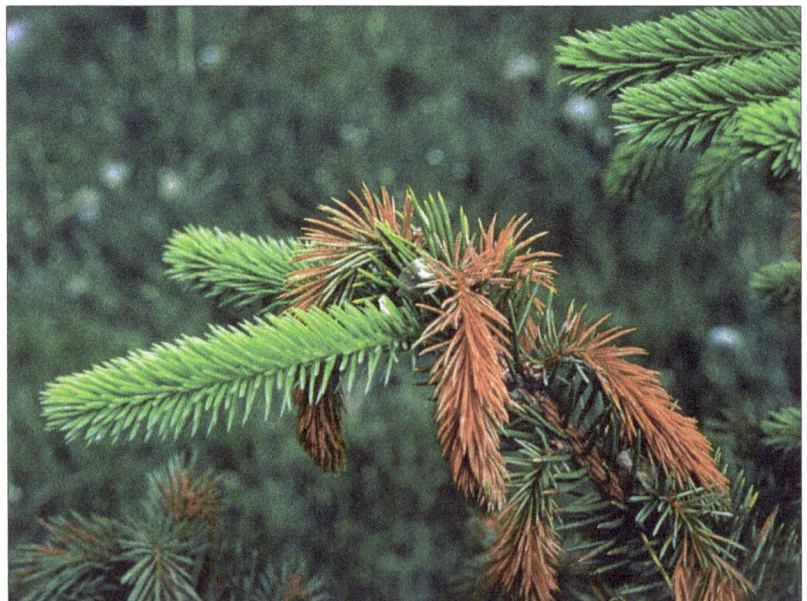

Frost-killed shoots on a spruce. (MN-DNR Archive, Bugwood.org)

Jack Pine Tip Beetle

Jack Pine Tip Beetle

Conophthorus resinosae

Hosts: Scotch, red, and jack pine

Importance: Feeding by larvae of this small beetle will kill the tips of terminal (top) and lateral (side) shoots. Dead shoot tips fall off the tree in late summer and autumn, producing an effect similar to shearing. Trees may be degraded by forking if a terminal shoot is killed and two or more lateral buds develop and become leaders. Serious damage is not common but can occur, mostly when susceptible pines are planted next to a jack pine stand.

Look For:

MAY TO OCTOBER

- *Yellow or reddish-brown shoot tips*, usually in the mid or upper canopy of the tree. The top 1 inch of the tip is killed and breaks off, leaving a flat stub.
- *A small pitch tube*—glob of pitch with a round hole in it—about ½ inch behind the bud.
- *A dark-brown beetle*, about 1/16 inch long, or several smaller, white larvae inside the injured tip. Peel away the bark on the shoot tip to find the insect.

Pests that cause similar symptoms: Pine shoot beetle, eastern pine shoot borer, and Diplodia shoot blight all kill shoot tips.

Monitoring and Control:
Inspect trees 2 years before harvest.

- Shear injured tips during routine trimming. Injured tips missed in shearing will probably fall off naturally.
- Prune excess leaders to prevent forking.

Next Crop:

- Avoid planting susceptible pines within 50 feet of a jack pine stand.

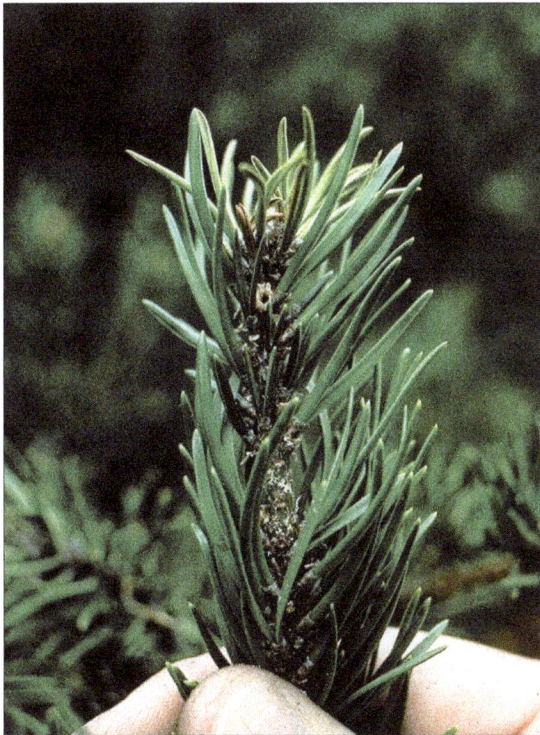

Pitch tube below the bud. (USFS - NCRS Archive)

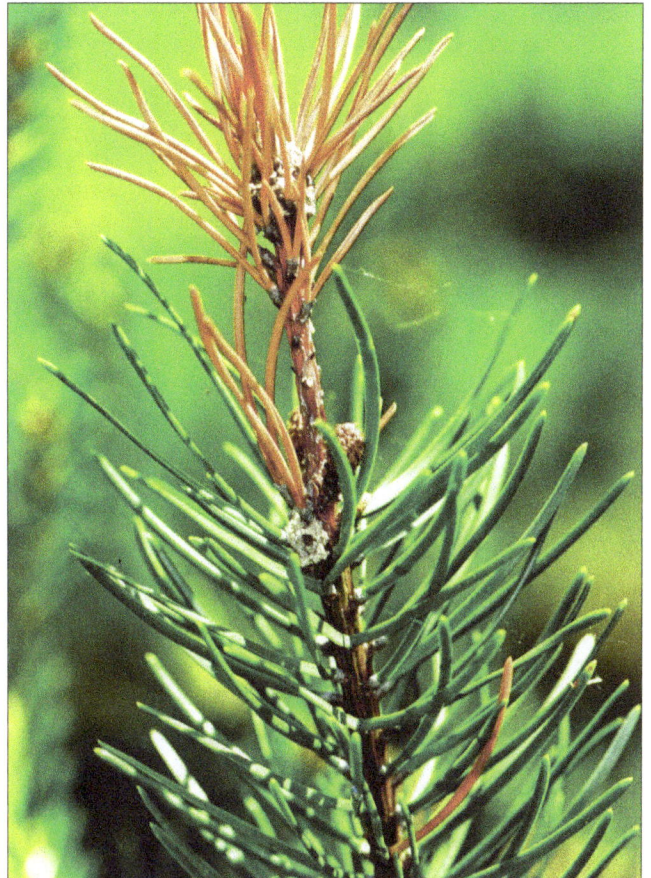

Flagged or dead branch tip. (S. Katovich, USFS, Bugwood.org)

Leucostoma (Cytospora) Canker

Leucostoma kunzei

Hosts: Spruces, especially Colorado blue and Norway

Importance: This disease usually affects trees older than 15 years that are stressed by drought, winter injury, or other diseases. Cankers degrade trees by killing foliage and branches. Stem cankers can eventually girdle and kill trees.

Look For:

- *Brown needles on lower branches*. Dead needles may drop off immediately or stay on the tree for up to a year. Leucostoma canker gradually kills lower branches first, and then the fungus spreads to upper branches.

- *Large, white patches of pitch* at canker sites. Cankers are hard to see if pitch is not present because the outer bark looks normal. Use a hand lens to find the tiny, black fruitbodies of the fungus in the bark beyond the canker.

- *Dead, brown areas of the inner bark* of affected branches that are visible when the outer bark is removed.

Factors that cause similar symptoms: Drought, Rhizosphaera needlecast of spruce

Biology: Spores ooze from the fruitbodies in droplets or threadlike masses during wet weather and are spread by rain, wind, and cultural activities, such as pruning. The fungus infects stressed trees through wounds on branches and stems.

Monitoring and Control: Inspect trees, especially those 10 to 15 years old or older. Look for brown needles on dead lower branches at any time of year. Treat individual trees as soon as you notice injury.

- Remove infected branches. Do not prune or shear infected trees during wet weather because spores released at this time may be carried from tree to tree on pruning tools.

- Improve tree vigor through cultural practices, such as fertilization and weed control.

- Sell trees growing on poor sites as soon as possible if Leucostoma canker has been a problem in your plantation. These trees are more likely to become infected than those growing on good sites.

- Avoid wounding trees because branch and stem wounds are entry points for the fungus that causes Leucostoma canker.

Next Crop:

- Do not plant susceptible species on poor sites.

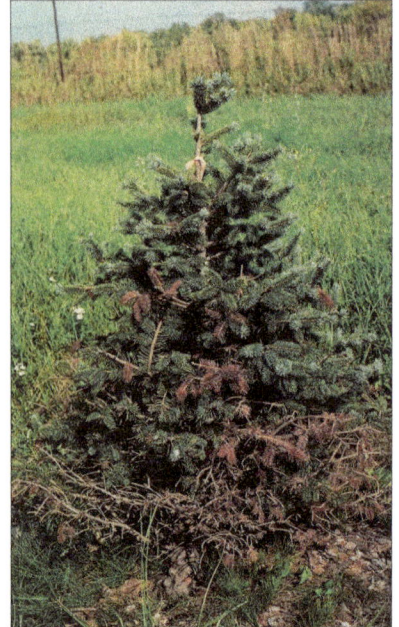

Scattered dead and dying branches. (USFS - NCRS Archive)

White patches of pitch on a branch canker. (USFS - NCRS Archive)

Nantucket Pine Tip Moth

Rhyacionia frustrana

Hosts: Scotch, Austrian, and red pine

Importance: The caterpillars of this moth kill and deform shoots on young trees. The trees may become bushy and misshapen and are therefore degraded as Christmas trees. This pest occurs mainly in the southern portions of the North Central Region.

Look For:

- *Dead or dying new shoots with expanded needles* anywhere on seedlings or trees.

MID-MAY TO AUTUMN

- *Brown to orange larvae*, up to ⅜ inch long, on or inside new shoots.
- *Small, tent-like webbing* on the surface of needles or at the base of shoots.

Pests that cause similar symptoms: Diplodia shoot blight, European pine shoot moth, pine shoot beetle

Biology: Pupae overwinter in hollowed-out shoots. Adult moths emerge and lay eggs on shoots in early spring when the weather warms. Newly hatched larvae feed on new, expanding shoots under small, tent-like webs, and then pupate 3 to 4 weeks later. One or more generations follow. Dry weather and poor soil conditions encourage population buildup.

Monitoring and Control: Examine trees of all ages, especially in nurseries and during the first 5 years after planting. Check trees closely in mid- to late April.

- If the attack is light (less than 5 percent of trees) and scattered, prune off and destroy the injured shoots.
- Shear when larvae are feeding to remove infested shoots.
- If there are attacks on more than 5 percent of the trees, thoroughly spray shoot tips with a registered insecticide between mid-May and mid-June (timing depends on latitude) to control young larvae before they conceal themselves. You may need to treat again between mid-July and late August to control additional generations of larvae.
- Avoid or limit the use of broad-spectrum insecticides that can kill beneficial natural enemies of other insect pests.

Caterpillar in a shoot. (D. Ross, Oregon State Univ., Bugwood.org)

Damaged, stunted, rounded tree. (E.R. Day, VA Polytechnic Institute and State Univ., Bugwood.org)

Pales Weevil

Hylobius pales

Hosts: Eastern white and Scotch pine, Douglas-fir, true firs, occasionally other pines and spruces

Importance: This insect is a chronic problem in Christmas tree plantations where periodic harvests leave many fresh pine stumps suitable for weevil breeding. The adult weevils feed on the bark of seedlings, shoots, and branches on older pines and other conifers. Seedlings often die. On older trees, feeding kills shoots, thereby thinning and degrading Christmas trees.

Look For:

JUNE TO AUGUST

- Dead seedlings.

- *Dead shoots* on larger trees.

- *Small, irregular patches of exposed wood* on the stems of seedlings or near the base of flagged (dead) shoots on larger trees. Pitch may ooze from the wounds, or the scars may be partially callused.

- *White, C-shaped larvae or pupae*, ½ inch long, beneath the bark of the large roots on fresh pine stumps. The larvae are in galleries that run parallel with the grain of the wood. Remove the bark from the large roots with a knife to look for the insects.

- *Reddish-brown to black, robust weevils*, ⅓ inch long, under the litter around live trees and stumps. Adults feed on trees during warm nights (temperature above 50 °F) from April to September.

Pests that cause similar symptoms: Diplodia shoot blight, Saratoga spittlebug, white pine blister rust, eastern pine shoot borer, jack pine tip beetle, and pine shoot beetle can kill shoots. Galleries created by wood borer larvae can result in holes that are apparent in the top of old stumps. White grubs (larvae) can kill seedlings.

Biology: Female weevils attracted by the odor of fresh pine resin in spring lay eggs in the inner bark of fresh stumps. Larvae feed in tunnels that may extend a few inches to several feet into the large roots during the summer. Adults emerge in late summer to early fall. They spend the day in the litter around live trees and move onto trees

Shoot tips killed by Pales weevil. (USFS - NCRS Archive)

at night to chew on the bark of shoots or seedlings. Later, adults move to the litter to overwinter.

Monitoring and Control:

Examine trees of all ages in June, especially where fresh stumps are present. Examine 50 or more trees scattered throughout the plantation. Treat the entire plantation if older trees average five or more flagged tips per tree. To survey for adults, put a sheet under a tree after dark and shake or rap the branches. Weevils will fall onto the sheet. Weevils can also be baited in spring by placing freshly cut pieces of pine stem on the ground near live trees. Look for adult weevils near or under the cut pieces during the day.

- See table 1 (page 22) for degree day information.

- Remove new stumps by early spring to eliminate breeding material.

- Alternatively, thoroughly drench the stumps and nearby soil with a registered insecticide

once between early April and mid-May to kill egg-laying adults. A similar drench in August will kill emerging adults.

- An alternative is to apply a registered insecticide to live trees between mid-August and mid-September when adults move onto trees to feed on shoots.

- When harvesting, leave one whorl of live, pest-free branches on the stump to keep it alive and therefore unattractive to the weevils.

Destroy these stumps within 3 years. Caution: Do not leave live branches that have needlecast diseases.

Next Crop:

- Delay replanting a cutover area for 2 years unless stumps are removed or treated to prevent weevil attack.

- Dip the above-ground portion of seedlings in a registered, residual insecticide before planting to prevent weevil feeding.

Feeding injury removes small patches of bark. (USFS - NCRS Archive)

Larvae and pupae at the top of a debarked stump. (USFS - NCRS Archive)

Pales weevil adult. (D. Shetlar, Ohio State University)

Phomopsis Canker

Phomopsis spp.

Hosts: Colorado blue, white, and Norway spruce

Importance: Species of *Phomopsis* cause a tip and shoot blight of nursery seedlings along with larger spruce in Christmas tree fields. In mixed-use fields where spruce are also dug for the landscape market, the fungus may kill the lower branches of balled and burlapped trees held at the farm or retail lot.

Look For:

- *Brown to purple needles and defoliation of live branches* anywhere on a tree.

- *Wilted or dead branch tips; eventually the branch completely dies.*

- *Resinous lesions and girdling cankers just under the bark* on older dead and dying branches, most often in the lower portion of trees. Removing bark may reveal cankers near the main stem, near dead or dying tips, or associated with branch nodes along the defoliated branches.

Pests that cause similar symptoms: Diplodia canker, Leucostoma canker, Rhizosphaera needlecast

Biology: *Phomopsis* is a known pathogen of spruce and should be managed in nurseries to reduce potential introduction to Christmas tree fields. At this time it is not known with certainty if *Phomopsis* is a primary pathogen singularly responsible for branch death symptoms and decline of large landscape trees. Two other factors may predispose trees to infection by *Phomopsis*: stressful environmental conditions or the presence of one or more other fungi that frequently can be found on needles and shoots of affected trees. It is thought that spores are released from fruitbodies during wet weather and splashed by rain to nearby needles and stems where infection takes place.

Monitoring and Control: Examine trees anytime during the year. Remove dead branches and severely affected trees; consider applying registered fungicides. Colorado blue spruce and white spruce appear more susceptible to branch death, and Norway spruce appears more tolerant. In each spruce species, there seems to be trees more susceptible to infection than others.

- Reduce moisture stress by controlling weed and grass competition.

- Use cultural practices to maintain high tree vigor.

- Do not shear during wet weather and avoid wounding trees.

- Remove sources of fungal inoculum such as dead, infected branches and shoots.

- Apply a registered fungicide to protect spruce during the infection period. Make the first application at bud break and then make a series of applications at 3-week intervals until the new growth is fully developed and hardened off.

Next Crop:

- Inspect planting stock carefully and plant disease-free stock.

Resinous lesion under the bark of an infected branch. (D. Fulbright, MSU)

Wilted branch tips on infected lower branches. (D. Fulbright, MSU)

Pine Grosbeak

Pine Grosbeak

Pinicola enucleator

Species Affected: Scotch pine; occasionally eastern white and red pine, and spruces

Importance: Pine grosbeaks feed on the buds of Christmas trees, stunting height growth and thinning crown foliage. This feeding causes dormant buds to develop into bushy clusters of shoots that deform and degrade trees. The extent of damage varies from year to year, depending on the number of birds and the supply of other food available during the winter months.

Look For:

WINTER

- *The pine grosbeak*—a robin-sized bird with a large, cone-shaped beak. Adult males are gray with a rosy-red coloring in the crown, rump, and breast. In females, these areas are suffused with yellow.

- *Buds missing from the topmost shoot and the upper branches* of trees.

- *Broken leaders on trees taller than 5 feet.*

MAY TO JULY

- *Bushy foliage in the upper part of the tree*, sprouting from lateral (side) buds that normally remain dormant.

Biology: During the winter months when their normal food supply is depleted, grosbeaks may migrate from northern forests south to areas that provide adequate food and shelter, such as Christmas tree plantations. Normally, this only occurs once every 4 or 5 years.

Monitoring and Control: The pine grosbeak is protected by the Federal Migratory Bird Treaty Act of 1918. Inspect trees within 2 or 3 years of harvest and protect them if you notice flocks of grosbeaks daily during the winter.

- If damage is common, install an electric noise broadcasting system to repel birds. Alter the kinds of sounds broadcasted every 3 to 5 days.

- Place a plastic mesh sleeve over the topmost shoot after the tree has become dormant, and remove it the following spring. This control technique is most practical for protecting high-value trees that will be harvested the next year.

- Shear damaged trees to help restore good form.

Adult male pine grosbeak. (J. Viola, Northeastern Univ., Bugwood.org)

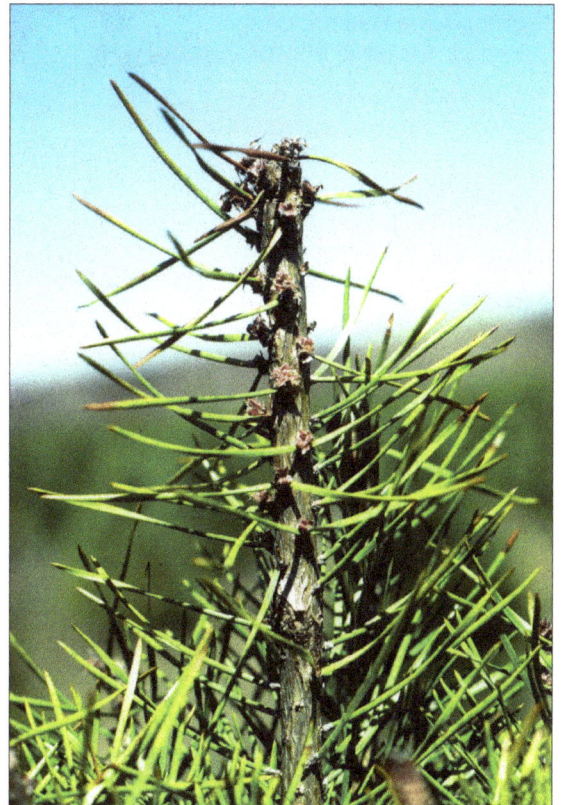

Feeding by grosbeaks removes the terminal buds. (USFS - NCRS Archive, Bugwood.org)

Pine Root Tip Weevil

Hylobius rhizophagus

Hosts: Scotch and red pine, eastern white pine if mixed with Scotch pine

Importance: Larvae (grubs) of this weevil feed on the root tips of host trees, preferring trees regenerated from stumps (tip-ups) because of the large root systems on these trees. Injured Scotch or red pines become discolored and may die. White pine may be attacked if grown near susceptible pines, but few will be killed. This weevil is found mostly in the northern Lake States on dry sandy sites.

Look For:

- *Flagged (deformed and discolored) shoots and branches* anywhere on the tree. Some trees may be dead.

- *Debarked, hollowed-out root ends* where root tips have been chewed off.

- *White, C-shaped larvae,* up to ½ inch long, which may be in the root.

Pests that cause similar symptoms: Diplodia shoot blight, pine spittlebug, Saratoga spittlebug, and Scleroderris canker. White grubs damage roots on seedlings, not larger trees.

Biology: Both larvae and adults overwinter underground. Overwintering adults (weevils) emerge in April and lay eggs mostly in June. Adults are nocturnal. Newly hatched larvae feed on pine rootlets and then tunnel into the main lateral (side) roots as they get larger. These larvae will overwinter.

Overwintering larvae feed on roots into early summer when they pupate, and new adults develop in late summer. They will overwinter and lay eggs the following spring.

Monitoring and Control: Diagnosis of this weevil is difficult. If you suspect pine root tip weevil, ask a pest specialist to confirm the diagnosis and recommend a treatment plan.

Next Crop:

- Remove any abandoned Scotch pine trees that are adjacent to areas planned for planting.

- Avoid growing trees from stumps (tip-ups).

- If the problem is serious, consider planting a nonhost species after harvest.

Debarked, hollowed-out root with a chewed-off tip. (USFS - NCRS Archive)

Flagged shoots and branches. (USFS - NCRS Archive)

Pine Shoot Beetle

Tomicus piniperda

Hosts: All pines, prefers Scotch pine

Importance: Pine shoot beetle is an exotic bark beetle. Larvae feed on the inner bark and cambium of recently killed, cut, or dying pines. Adults, however, feed in the shoots of live pine trees during the summer; this feeding can reduce the aesthetic quality of trees. In most cases, however, pine shoot beetle is important because it is a quarantine pest. Quarantines regulate shipments of pine Christmas trees from States and counties known to be infested.

Look For:

- *Reddish boring dust* on the bark of recently cut pine trees or stumps early in spring when parent beetles are colonizing brood material.

- *Egg and larval galleries under the bark* of recently cut pine trees or stumps that have been colonized by parent beetles. Egg galleries run parallel with the grain of the wood and have a slight bend at the end. Larvae feed in tunnels that are roughly perpendicular to the egg gallery.

JUNE TO DECEMBER

- *Attacked shoots on live pine trees with a round hole,* often surrounded by a small glob of pitch. Beetles feed in tunnels down the center pith of the shoots. Tunnels are hollow and are not filled with frass. Two to five tunnels may be found on a single shoot. Attacked shoots eventually die, break off, and drop to the ground.

Pests that cause similar symptoms: Other bark beetles breed and develop in the inner bark of recently cut pine trees, stumps, and logs. Shoots killed by Diplodia shoot blight will not have tunnels. Shoots attacked by eastern pine shoot borer and European pine shoot moth will have fine, sawdust-like frass packed into the tunnel. Jack pine tip beetle constructs hollow tunnels but only in the tips of shoots.

Biology: Adult beetles overwinter in a niche in the bark at the base of live pine trees. In late winter or early spring, adult beetles fly to recently cut or dying pine trees. Adults bore into the inner bark and mate, and each female constructs an egg gallery that runs parallel with the grain of the wood. Eggs are laid along the sides of the gallery and hatch within a few weeks. Larvae feed in the inner bark for 6 to 10 weeks and pupate. New adults emerge in early summer. Adults feed in current-year or 1-year-old

Cut trees and other brood material should be burned or chipped. (D. McCullough, MSU)

shoots on live pine trees until October or November. After a few hard frosts, beetles move to the base of trees to overwinter.

Monitoring and Control: If you plan to ship pine trees outside of your county, contact your State regulatory agency for current information on quarantines and regulations. Check live trees throughout the summer and fall for evidence of shoot feeding. Clip off discolored or dead

shoots and split them lengthwise to look for a black or reddish-brown beetle feeding in a hollow tunnel.

- See table 1 (page 22) for degree day information.

- Reduce availability of dying pine trees, or pine trees or stumps cut within the past year. Chip or burn culled or unsold trees by late spring.

- Cut stumps as low to the ground as possible. Stumps

can also be sprayed with an approved insecticide in late spring before new adult beetles emerge.

- Freshly cut pine trees or logs can be set along edges of fields in early spring to attract parent beetles. These "trap trees" must be collected and destroyed before the progeny beetles can emerge.

- Spraying foliage with a registered insecticide can help, but is rarely 100 percent effective. Time the spray to coincide with emergence of the new generation of beetles.

Next Crop:

- Practice good sanitation in fields to prevent populations of pine shoot beetle from building to damaging levels.

Flagged, infested shoots. (E.R. Hoebeke, Cornell Univ., Bugwood.org)

Pitch tubes. (E.R. Hoebeke, Cornell Univ., Bugwood.org)

Pine Spittlebug

Aphrophora parallela

Hosts: Scotch, Austrian, and eastern white pine; all spruces and firs

Importance: Spittlebug nymphs and adults suck sap from shoots of Christmas trees. Unless abundant, they seldom do more than flag (discolor and deform) an occasional branch tip. However, the fungal pathogen *Diplodia pinea* may invade weakened pines through spittlebug feeding wounds, and shoot blight may heavily flag or kill trees.

Look For:

- *Flagged shoot tips* anywhere on the tree (especially on pines). Foliage may look sooty and glisten as if lacquered.

MAY TO EARLY JULY

- *Frothy, white spittle masses* on shoots or trunk.

- *One or more creamy yellow to black nymphs*, up to ¼ inch long, within spittle masses on a tree.

- *Shoot blight caused by Diplodia* on pines.

MID-JUNE TO MID-SEPTEMBER

- *Oval-shaped adults*, about ⅓ inch long, on needles or branches. They jump when approached or touched.

Pests that cause similar symptoms: Jack pine tip beetle, pales weevil, pine shoot beetle, pine root tip weevil, Saratoga spittlebug, Scleroderris canker, and Diplodia shoot blight can cause flagged shoots. Sooty, lacquered foliage could be due to aphids or pine tortoise scale.

Biology: Nymphs hatch in May from eggs laid under the bark of shoots. For the next 6 to 7 weeks, they feed on the tree's sap and produce the characteristic frothy spittle masses from partially digested sap. Black sooty mold grows on the sugary sap splashed from the spittle masses. Adults appearing in July also suck the tree's sap, but form no spittle masses. The Diplodia shoot blight fungus enters the feeding wounds and causes shoot tips to turn brown.

Monitoring and Control: Examine trees of all ages from May through June. A few scattered spittle masses need no treatment if trees are otherwise healthy. If insects seem abundant—as if trees are partially coated with "snow"— look for flagging in late summer and early fall. If trees are flagged, or if Diplodia shoot blight is also present, treat the entire plantation the next summer.

- Apply a registered insecticide in early to mid-July to control the adults. To determine spray date, start examining spittle masses in early July, and spray when 90 percent of them are empty. Manage Diplodia shoot blight if needed.

Next Crop:

- Select appropriate species for site conditions. Trees stressed by drought, poor growing conditions, or other factors are more susceptible to spittlebug and shoot blight injury.

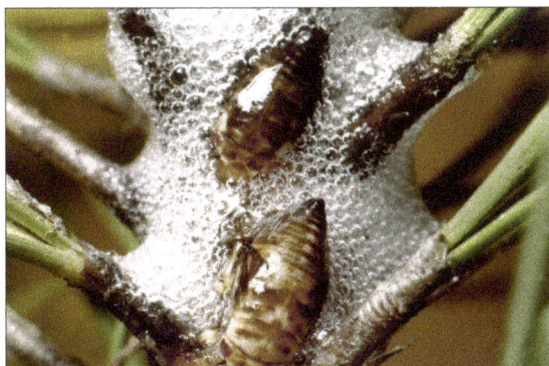

Immature nymphs in spittlemass. (Clemson Univ. - USDA Cooperative Extension Slide Series, Bugwood.org)

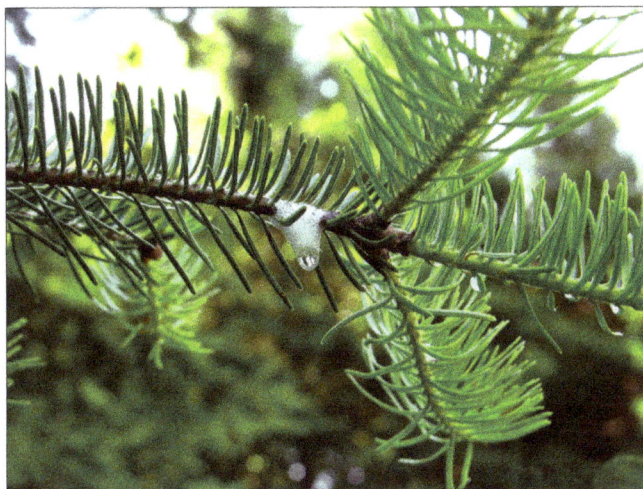

Spittlemass on balsam fir. (R.S. Kelley, VT Dept. of Forests, Parks and Recreation, Bugwood.org)

Pine Tortoise Scale

Toumeyella parvicornis

Hosts: Scotch, Austrian, red, and jack pine

Importance: Scales feed on sap of the shoots and branches of pines. Needles and branches of infested trees become coated with black sooty mold, which grows on honeydew excreted by the scales. Scale feeding and black sooty mold cause needles and shoots to become discolored. Infestations can leave trees unfit for sale and can kill branches or small trees.

Look For:

- *Discolored needles and dying shoots or branches*, particularly in the lower canopy. Needles may glisten as if lacquered or appear black and sooty. Bees, wasps, or ants, which are attracted to the honeydew, may be present.

- *Reddish-brown, mottled, helmet-shaped scales*, up to ¼ inch in diameter, on the bark of injured shoots and branches.

Pests that cause similar symptoms: Sooty mold can occur on trees infested with aphids.

Biology: Immature female scales overwinter on the bark of shoots and branches. They begin feeding in spring, mature in early summer, then lay eggs. Each female can produce roughly 500 eggs. In June or July, tiny, pinkish crawlers (nymphs) hatch and crawl out from beneath the female scale. Crawlers move about on the tree until they settle and begin to feed. Feeding scales secrete a sugary waste product called honeydew that coats the nearby needles and branches. Needles may appear shiny at first, but black sooty mold will grow on the honeydew, which turns the needles and shoots black. Needles, shoots, and entire branches can be killed. Adult and larval lady beetles, along with other predators, frequently feed on scale eggs and crawlers and can usually control light infestations. Bees, wasps, and ants will feed on the sugary honeydew. Ants often protect the scales from predators, which can lead to heavy infestations and tree damage.

Monitoring and Control: Examine trees of all ages from May to June, looking for sooty needles. Treat individual infested trees. Pay special attention to trees that are ready for harvest.

- See table 1 (page 22) for degree day information.

- Cut and burn heavily infested trees before eggs hatch around mid-June.

- Spray infested trees with horticultural oil or with a registered insecticide between mid-June and mid-July when crawlers are emerging. Use a hand lens to check for the tiny crawlers on the undersides of mature scales in mid-June. Spray when roughly half of these crawlers have emerged. If timing is incorrect, a second treatment may be needed.

- Avoid spraying chemical insecticides when lady beetles or their larvae are present.

- Control mound ants that protect scales from predators.

Next Crop:

- Do not plant susceptible pines next to a scale-infested stand or windbreak. Avoid Jack pine, a common host.

Pine tortoise scales on sooty bark. (USFS - NCRS Archive)

These white lady beetle larvae are predators of the scale crawlers. (B. Bishop, Concordia College, MN)

Saratoga Spittlebug

Aphrophora saratogensis

Hosts: Scotch and red pine; occasionally eastern white pine, Fraser fir, and balsam fir

Alternate Hosts: Sweetfern, brambles (raspberry and blackberry), hawkweed

Importance: Spittlebug adults feed on shoots of conifers and can discolor foliage, stunt or kill branches, and leave trees unfit for Christmas tree sale. Heavy feeding for 2 or 3 years can kill branches and trees. Saratoga spittlebug can be very damaging if the alternate hosts are nearby and abundant, especially sweetfern.

Look For:

- *Reddish or reddish-brown (flagged) branches*, particularly in the upper part of the tree. Severely injured trees turn brown and die.

- *Tan or brownish flecks on the wood* under the bark of older portions of branches. Peel bark away with a knife to see the flecks.

- *Dense ground cover of sweetfern or other alternate hosts*.

MID-MAY TO EARLY JULY

- *Frothy masses of spittle-like bubbles at the base of sweetfern or other alternate hosts*.

- *Small, red and black or chestnut-brown insects* in these spittle masses.

LATE JUNE TO SEPTEMBER

- *Tan and white, boat-shaped insects*, ⅓ inch long, on the tree. Each has a white, arrow-shaped marking on its front end. Spittlebug adults will jump away when disturbed.

Pests that cause similar symptoms: Diplodia shoot blight, pales weevil, jack pine tip beetle, pine shoot beetle, pine root tip weevil, Sirococcus shoot blight, and Scleroderris canker on pine can all cause flagged branches or shoots. Meadow spittlebug (green insects in spittle masses in the tops of weeds and grasses) is not a Christmas tree pest. Pine spittlebug forms a spittle mass on the tree, not on an alternate host.

Biology: Nymphs hatch in spring and drop from trees to feed on alternate host plants. They cover themselves with the characteristic spittle mass and can be found on the stems of alternate hosts often right at the ground line. In late June or early July, adults return to the trees to suck sap from the shoots and lay eggs. Adult feeding scars the inner bark and wood. This damage can kill twigs and larger branches.

Monitoring and Control: Examine sites before planting and check between rows of young trees after planting. Look for and eliminate pockets of sweetfern and other alternate hosts if they occupy more than 20 percent of the open ground cover. If branch flagging is observed, randomly select five host trees scattered throughout

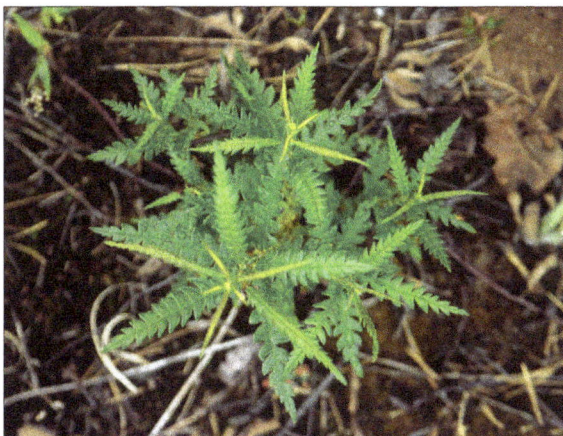

Sweetfern is the alternate host for the Saratoga spittlebug. (S. Katovich, USFS, Bugwood.org)

Spittlemass at the base of sweetfern. (USFS - NCRS Archive)

the plantation. Scrape the bark off the 2-year-old shoot portion of any branch in the upper half of each tree and look for flecks (feeding scars) on the wood. Consider treating adult spittlebugs if there are more than 20 flecks per 4 inches of branch length.

- Limit sweetfern and other alternate hosts to less than 20 percent of the ground cover within 20 feet of the trees.

- Apply a registered insecticide to the trees in early to mid-July to kill emerging adults before they lay eggs. The best time to treat trees is when 90 percent of the spittle masses on sweetfern are empty.

Next Crop:

- Eliminate alternate hosts before planting trees. Limit alternate hosts throughout the growing cycle to less than 20 percent of the ground cover.

- Avoid very dry, sandy soils that are prone to drought stress. These areas often have sweetfern growing on them.

Tan flecks under the bark are feeding wounds. (G. Simmons)

Adult Saratoga spittlebug. (USFS – NA Archive, Bugwood.org)

Flagged branches. (USFS - NCRS Archive)

Scleroderris Canker

Gremmeniella abietina
(Syn. *Scleroderris lagerbergii*),
G. balsamea

Hosts: All pines; occasionally spruces, firs, and Douglas-fir

Importance: Scleroderris canker is caused by several species and races of the fungus *Gremmeniella.* The European strain of *G. abietina* kills trees of all ages, causing extensive losses in plantations. The North American strain of *G. abietina* only kills trees less than 6 feet tall and can seriously damage Christmas trees during the first 5 years after planting. Only the North American strain is present in the Upper Midwest. Balsam fir is infected by *G. balsamea*.

Look For:

- *Cankers and resinous lesions* that are inconspicuous *sunken areas* on the stem and branches.

- *Green discoloration of the wood beneath the bark* of dead branches.

MAY TO JUNE

- *Orange discoloration at the bases of needles*, usually on the lower 3 feet of the tree. These needles fall off easily.

- *Dead buds on shoots* that have discolored needles.

JULY TO NOVEMBER

- *Brown needles and branch tips*. Needles fall off when touched.

DECEMBER TO APRIL

Dead branch tips with no needles.

Pests that cause similar symptoms: Diplodia shoot blight and canker, drought, pine root tip weevil, Saratoga spittlebug

Biology: Branch tips usually become infected in May and June, but infection sometimes takes place from February through November. Spores are windblown or rain splashed from infected trees. The disease is most severe in areas where summers are cool and winters are long.

Monitoring and Control:
Examine trees of all ages in May or June when orange needle discoloration is most obvious. Check the lower branches of trees growing in depressions, and remove infected branches. Continue to check branches each year, especially if this disease is present in natural stands within ¼ mile.

- Remove all infected branches.

- Do not shear infected foliage during wet weather because spores released at this time can be carried from tree to tree on shearing tools. Sterilize tools after shearing infected trees by dipping in disinfectant for at least 15 seconds.

- Shear healthy trees first so spores will not be carried from infected trees to healthy ones.

- Chemical controls are available to protect nursery stock, but may be too expensive for plantation use because of the large number of applications required.

Next Crop:

- Plant resistant species in areas where Scleroderris canker is present.

- Avoid planting in frost pockets that favor disease development.

Green discoloration of a killed branch. (USFS - NCRS Archive, Bugwood.org)

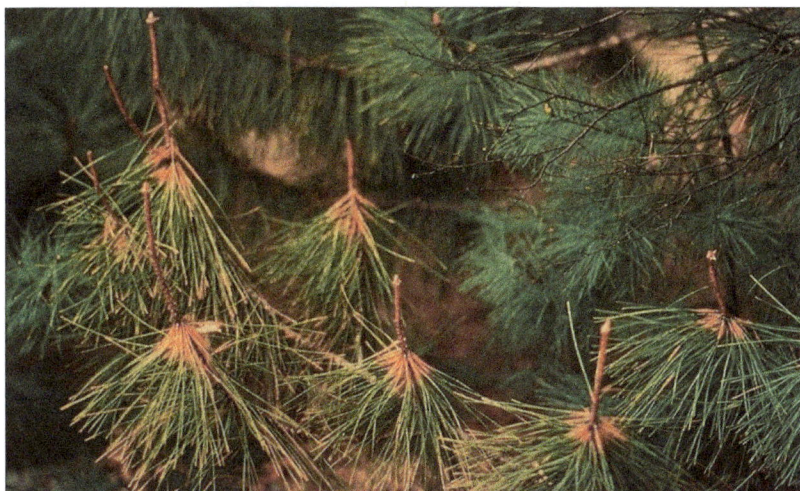

Orange discoloration at the bases of infected needles. (USFS - NCRS Archive, Bugwood.org)

Sirococcus Shoot Blight

Sirococcus conigenus

Hosts: Red and Scotch pines, Colorado blue spruce and white spruce on rare occasions

Importance: This fungus causes a shoot blight that can kill seedlings of several conifer species. Current-season shoots and occasionally 1-year-old branches are killed, disfiguring plantation and landscape trees.

Look For:

- *Lesions on succulent shoots and, later, drooping shoots* resembling shepherds crooks in spring and early summer.

- *One-year-old dead shoots* with black fruitbodies on needles and the stem.

Pests that cause similar symptoms: Diplodia shoot blight, Saratoga spittlebug on Scotch and red pine

Biology: The fungus is favored by cool, wet weather and low light conditions most often found in the lower portion of trees. Spores produced in fruitbodies in previously killed needles and shoots are released in wet weather by rain splash during May and June and infect nearby young needles. The fungus grows into elongating shoots, eventually killing them, and overwinters in the dead tissues, completing its 1-year life cycle.

Monitoring and Control: Inspect trees of all ages during spring and early summer for the drooping of new shoots or shoots killed the previous season.

- Remove drooping or killed shoots during dry weather to reduce sources of fungal spores.

- Do not shear trees during wet weather to avoid spreading the fungus.

Next Crop:

- Buy disease-free planting stock.

- Avoid planting trees in low areas where high humidity favors the disease.

- Avoid planting spruce or pines near older, larger trees of the same species that may harbor the fungus.

Killed current-year shoots. (J. O'Brien, USFS, Bugwood.org)

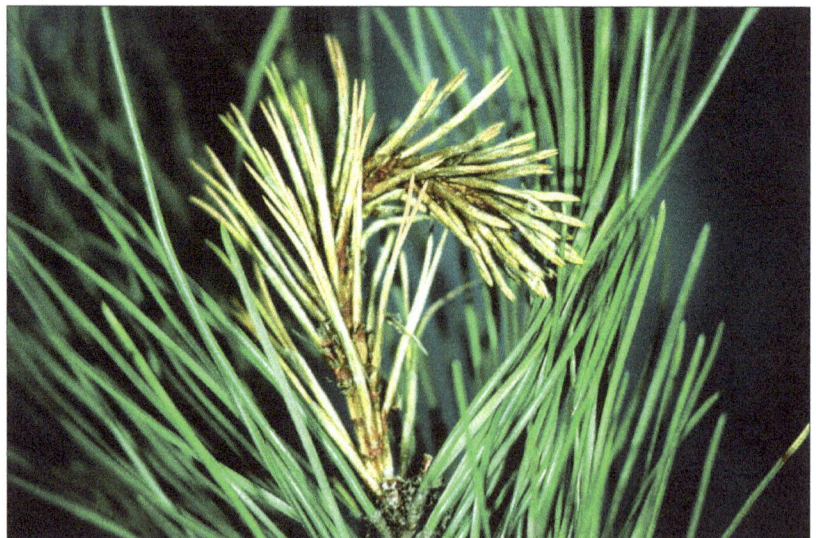

Curled, infected current-year shoot. (USFS - NCRS Archive, Bugwood.org)

Spruce Bud Scale

Physokermes piceae

Hosts: Can infest all spruces, although Norway spruce is preferred

Importance: This scale sucks fluids from the shoots and branches of Christmas trees, but usually does not damage trees. Heavy attacks, however, may kill a few trees or stunt new shoots, leaving trees unfit for sale.

Look For:

- *Discolored needles and dying shoots*, particularly on the lower branches.
- *Dusty red-brown, globe-like scales*, up to 3/16 inch in diameter, that look like abnormal buds at the bases of current shoots.
- *Black sooty mold* on infested trees.

Biology: Immature female scales overwinter on branches where they cluster around terminal buds with their straw-like mouthparts embedded into twigs. Feeding resumes in the spring. Scales mature by early summer into plump, brown bumps that can be found at the base of twigs that arise from terminal buds. Eggs laid under the scale body hatch in a few weeks, and crawlers migrate towards the tips of twigs where they settle and feed until winter. There is one generation a year. Lady beetles frequently feed on the crawlers.

Monitoring and Control: Examine trees the year before harvest. Lady beetles and lacewings will usually control light infestations. However, consider control options if predators are not effective or if you notice scales on trees that are ready for harvest.

- See table 1 (page 22) for degree day information.
- Cut and burn heavily infested trees before mid-June to reduce spread of crawlers.
- Treat trees with a dormant oil before buds break in spring to kill immature scales.
- Alternatively, spray infested trees once with a registered insecticide between mid-June and mid-July to kill emerging crawlers. Use a hand lens to check for pinkish eggs or small crawling insects on the undersides of scales in mid-June. The best time to spray is when almost half of these crawlers have emerged. If timing is incorrect, a second spray may be needed.

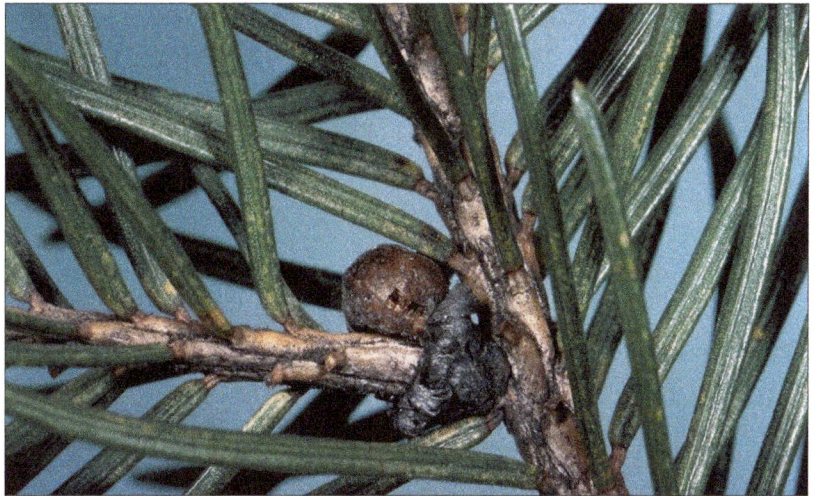

Spruce bud scale looks like a bud. (G. Simmons, MSU)

Mature adult scales. (S. Katovich, USFS, Bugwood.org)

White Pine Blister Rust

Cronartium ribicola

Host: Eastern white pine (all 5-needle pines)

Alternate Hosts: Gooseberry, currant, *Ribes* spp.

Reddish-brown needles above a main-stem canker. (USFS - NCRS Archive)

Importance: White pine blister rust causes cankers that kill branches and lower the market value of Christmas trees. Cankers on the trunk can girdle and kill trees.

Look For (on pine):

- *Patches of brown bark with yellow borders* developing during the first year of infection.
- *Spindle-shaped swellings* that develop on the branches or trunk during the second year after infection.
- *Resin flow and rodent feeding* on older branch and trunk cankers.
- *Reddish-brown needles* on dead branches and tree tops above trunk-girdling cankers.

MAY

- *Cream-colored blisters* pushing through the diseased bark of cankers. These blisters break open and release powdery, orange-yellow spores that infect gooseberry and currant plants.

JUNE TO JULY

- *Yellow-brown blisters* on the canker that produce a sticky, orange fluid that later hardens and turns black. Rodents are attracted to cankers at this stage and evidence of feeding is common.

Look For (on gooseberry and currant):

JUNE TO AUGUST

- *Yellow-orange spores* on the undersides of leaves.

AUGUST TO OCTOBER

- *Brown, hair-like fungal growths* on the undersides of leaves. The spore stage that infects pines, which completes the disease cycle, is produced on these structures.

Pests that cause similar symptoms: Armillaria root disease, vole damage, and pine root collar weevil can kill scattered trees. Pales weevil can cause branch flagging.

Biology: This fungus needs both pine and an alternate host to complete its long, complicated life cycle. It spreads from pine to gooseberry or currant, but cannot spread from pine to pine. The disease is usually most severe in northerly regions on sites where the weather is cool and moist in August and September.

Monitoring and Control:
Inspect trees throughout the year. Look for branch flagging, orange blisters, or pitchy cankers on branches or trunks. If trees are infected, remove cankers if they are on branches and remove alternate hosts before August.

- When shearing Christmas trees, prune all flagged (brown) branches that have cankers. This prevents the fungus from entering the trunk and killing the tree.

- Eliminate trees with trunk cankers; they will never produce quality trees.

- Pull out or kill alternate hosts with a registered herbicide.

Next Crop:
- Avoid planting white pine where alternate hosts are abundant, especially in high-hazard, northern areas. If you do plant in these areas, remove alternate hosts to reduce the likelihood of infection.

- Avoid planting white pine in areas where cool air collects, such as at the base of a slope or in a depression.

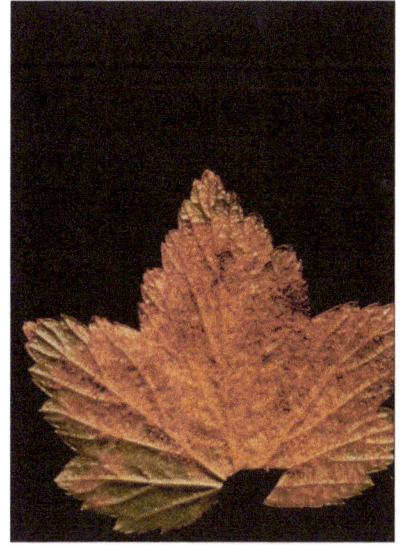
Brown, hairlike fungal growth on the underside of a Ribes leaf. (USFS - NCRS Archive)

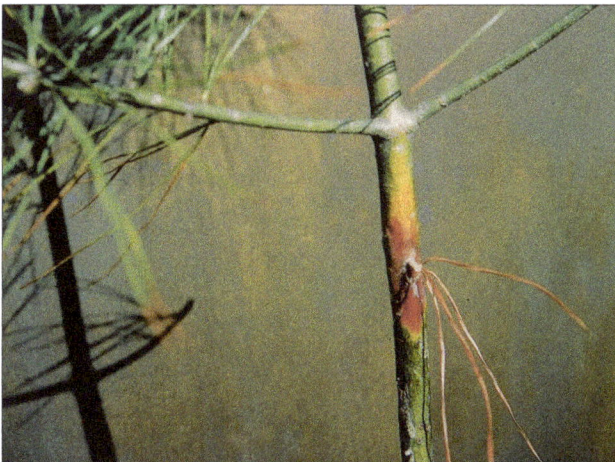
Young canker and killed needles. (USFS - NCRS Archive)

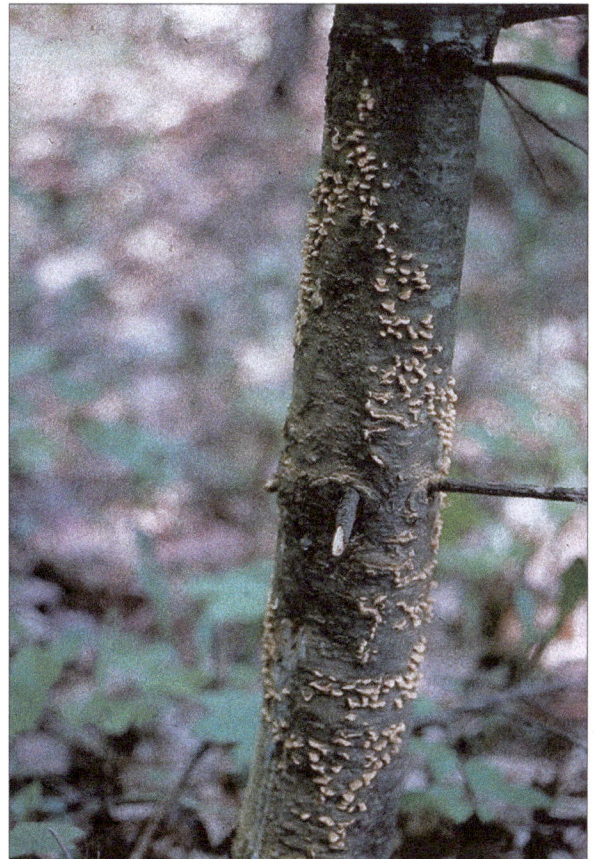
Fungal blisters pushing through diseased bark. (USFS - NCRS Archive)

White Pine Weevil

Pissodes strobi

Hosts: Eastern white and Scotch pines; spruces, especially Norway spruce; other pines are also susceptible

Importance: The larvae of this common pest deform and degrade Christmas trees by killing the terminal leader and the top 2 to 4 years of growth. Damage will delay harvest for 1 to 3 years until those trees recover enough to be suitable for sale.

Look For:

- *Dead or dying terminal leader* (topmost shoot on the main stem) curled into the shape of a shepherd's crook. Lateral (side) branches on the upper whorls may also die.

MARCH TO APRIL

- *Small, round holes or pitch flow on the terminal leader* where adult weevil is feeding or laying eggs.

JUNE TO AUGUST

- *Slightly curved, white larvae,* up to ¼ inch long, under the bark or in the wood of the damaged terminal.

- *Clumps of fine, white slivers of wood under the bark* in late summer. These chip cocoons may contain white pupae or brownish weevils, ¼ inch long.

Pests that cause similar symptoms: White pine blister rust can kill terminal portions of trees if a canker has formed on the main stem. Eastern pine shoot borer infests both

White pine weevil adult and chip cocoon. (J. Hanson, USFS, Bugwood.org)

White pine weevil larva. (J. O'Brien, USFS, Bugwood.org)

terminals and laterals. Frost can kill terminals, though frost would generally damage other new growth as well.

Biology: On warm spring days overwintering adults move from the ground to the tree tops to mate and lay eggs in holes they chew into the bark, just below the terminal buds on the previous year's leader. The eggs soon hatch and larvae bore downward under the bark, eventually girdling the top of the stem and killing the new terminal growth that has started to

expand above the feeding larvae. Larvae pupate in wood fiber cocoons, called chip cocoons, and emerge as adults from late July to late August. Adults feed on the bark of small branches before dropping to the litter to overwinter.

Monitoring and Control: Begin checking for dying and dead terminals in late June, and concentrate on trees that will be harvested in 3 or 4 years. Treat the entire plantation when injury becomes too severe to correct with pruning.

- See table 1 (page 22) for degree day information.

- Prune out and burn infested leaders before mid-July to kill the insects. Cut back all but one live lateral (side) shoot by at least half their length to maintain single-stem dominance.

- Spray only the terminal leader of trees with a registered insecticide as soon as the weather warms to control egg-laying weevils. Eggs are usually laid in early May in the Lake States and in April in the Central States. A second spray between mid-August and late September may be needed to control newly emerged adults. Apply the second spray to the upper half of the tree canopy.

Next Crop:

- Avoid planting highly susceptible white pine and Norway spruce. If you do plant these species, isolate them from less susceptible pines and spruces.

- Plant resistant varieties of Scotch pines, such as the Swedish variety, if available.

- If practical, remove eastern white pine, jack pine, and Norway spruce growing in and around plantations before planting.

Dead terminal on blue spruce. (D. Herms, The Ohio State Univ., Bugwood.org)

Dead terminal on white pine. (E.R. Day, VA Polytechnic Institute and State Univ., Bugwood.org)

Shoot/Branch Galls

Irregular or globelike swellings occur on shoot, branch, or mainstem. You may also find small pitch blisters in branch crotches that look like galls.

Cedar-Apple Rust

Gymnosporangium juniperi-virginianae

Host: Eastern redcedar

Alternate Hosts: Apple and crabapple trees

Importance: This rust causes unsightly galls (globe-like swellings) to form on infected eastern redcedar, resulting in twig dieback. Numerous galls may slow tree growth and kill seedlings.

Look For:

- *Brown, warty galls*, ½ to 2 inches in diameter, on twigs of redcedar.

MAY TO JUNE

- *Yellow-orange, jelly-like fingers growing from galls* on redcedar, especially during rainy weather.

JULY TO SEPTEMBER

- *Orange-yellow lesions* on leaves and fruit of nearby apple and crabapple trees.

Biology: This disease has a 2-year life cycle that alternates between redcedar and apple or crabapple trees. After warm spring rains, cedar-apple rust spores are produced in the yellow-orange fingers that erupt from round, woody galls on redcedar twigs. These spores spread to nearby apple and crabapple trees where they cause orange-yellow lesions on leaves and fruits. In summer and early fall, another type of spore produced on the apple and crabapple trees infects nearby redcedar and causes new galls to form that produce spores the following year, completing the life cycle.

Monitoring and Control: In spring, look for galls on redcedar of all ages. If galls are too numerous to hand clip and are killing seedlings or making older trees unsalable, consider treating galls with fungicides.

- Clip off galls on redcedar.
- If practical, remove nearby apple and crabapple trees to reduce infection of redcedar.
- Apply a registered fungicide to the orange, jelly-like galls on redcedar once during the spring and/or spray redcedar foliage three times, once every 2 weeks, beginning in midsummer.

Next Crop:

- Avoid planting redcedar near apple and crabapple trees.
- Plant juniper species and cultivars resistant to the disease if it is not practical to remove alternate hosts.

Rust symptoms on apple leaf. (Clemson Univ. - USDA Cooperative Extension Slide Series, Bugwood.org)

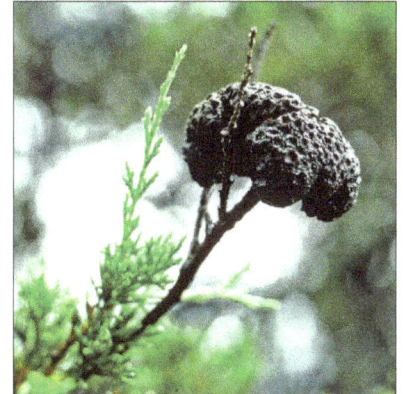

Brown, warty gall on redcedar. (USFS - NA Archive, Bugwood.org)

Jelly-like fingers growing from galls on redcedar. (J. O'Brien, USFS, Bugwood.org)

Cooley Spruce Gall Adelgid

Adelges cooleyi

Hosts: Colorado blue spruce and Douglas-fir

Importance: On blue spruce, feeding by adelgid nymphs will cause long, curved galls to form at the ends of the current-year shoots. Galls affect the appearance of trees and lower their value. On Douglas-fir, adelgid feeding discolors and distorts needles, and heavy infestations can reduce growth. No galls are produced on Douglas-fir.

Look For (on blue spruce):

- *Galls: cone-like and green, purple, or reddish-brown swellings*, 2 to 3 inches long, on the ends of current-year shoots.

Look For (on Douglas-fir):

- *Small, cottony balls* on needles; usually heaviest on the underside of needles on shoots in the mid or lower canopy.

- *Yellow spots on bent or curled needles*, caused by adelgid feeding.

Biology: These adelgids can complete their life cycle on blue spruce or Douglas-fir. On *spruce*, immature females overwinter near the base of terminal buds. As buds begin to swell in spring, females mature and lay eggs under a mass of white, cottony wax. Nymphs hatch and feed at the base of new expanding needles. This feeding causes the tree to produce galls that enclose and protect the adelgids. In midsummer, galls dry out, turn brown, and open, allowing the adelgids to escape. Adelgids can either continue their life cycle on blue spruce or fly to Douglas-fir to lay eggs.

On *Douglas-fir*, adelgids feed within tiny, white, cottony balls, but their feeding does not cause galls. Adelgids, which may overwinter as eggs or nymphs on needles, begin feeding early in spring. A second generation of eggs may hatch in late June or early July. These nymphs can feed on current-year needles as well as older needles.

Monitoring and Control:

- See table 1 (page 22) for degree day information.

On blue spruce: Clip off and destroy green galls before they turn brown and open in July. Brown galls can be removed in late summer or fall to improve tree appearance, but this will not provide any adelgid control.

- Chip or burn heavily affected trees.

- If galls are too numerous to hand clip, treat the plantation. Trees can be sprayed with a registered insecticide just before spruce buds break in spring or in the fall. A horticultural oil or dormant oil can be used but may temporarily affect tree color. Avoid spraying trees at temperatures below 55-60°F or above 75-80°F.

On Douglas-fir: Monitor trees of all ages throughout the growing season. If you find small, cottony balls on the undersides of needles, treat the entire plantation.

- Apply a registered insecticide, horticultural oil, or dormant oil to trees in fall (October) or just before Douglas-fir buds break in spring. Avoid spraying trees when the temperature is less than 55-60°F.

Cottony balls on Douglas-fir needles. (S. Katovich, USFS, Bugwood.org)

• An application in late June to mid-July may be needed if adelgid nymphs are abundant. Look for tiny, dark young nymphs on current-year needles as well as older adelgids encased in the white, cottony balls.

Next Crop:

• Avoid planting Colorado blue spruce and Douglas-fir near each other. Feeding may be more severe when adelgids can feed on both hosts.

Older, reddish-brown gall on blue spruce. (R.S. Kelley, VT Dept. of Forests, Parks and Recreation, Bugwood.org)

Green to purple galls on blue spruce. (W. Cranshaw, Colorado State Univ., Bugwood.org)

Eastern Gall Rust

Cronartium quercuum

Western Gall Rust

Peridermium harknessii
Syn. *Endocronartium harknessii*

Host: Scotch pine; also jack pine

Alternate Hosts:

Eastern gall rust: oak species
Western gall rust: none

Importance: Rust galls on stems slow growth and gradually kill older trees. Young seedlings are girdled and killed quickly. Rust galls on branches kill individual branches, but not trees.

Look For:

- *Galls: globe-like or spindle-shaped swellings* on stems or branches.

- *Red needles* on recently killed branches.

APRIL TO JUNE

- *Cream-colored blisters filled with orange spores* on the surface of galls.

Biology: Windborne spores of pine-pine (western) gall rust spread directly from pine to pine. Pine-oak (eastern) gall rust completes part of its life cycle on oak leaves.

Monitoring and Control:
Randomly select trees and look for branch and stem galls at any time during the year. If trees up to 7 years old average more than three galls per tree, consider treating the entire plantation.

- Destroy and remove oak alternate hosts within ¼ mile of plantations, if feasible.

- Control pine-pine gall rust by removing galls from trees before they produce spores that can infect other pines.

- Control rusts in nurseries by applying a registered preventive fungicide. Have a pest specialist identify the rust and prescribe the best treatment and timing for your area.

Next Crop:

- Before planting, inspect seedlings for stem swellings caused by rust infection. Do not plant diseased seedlings.

- Replant failed plantations with a nonhost species.

Stem gall covered with spores. (USFS - NCRS Archive)

Globe-like rust gall. (USFS - NCRS Archive)

Spindle-shaped rust galls on seedlings. (USFS - NCRS Archive)

Eastern Spruce Gall Adelgid

Eastern Spruce Gall Adelgid

Adelges abietis

Hosts: White, Black Hills, Norway, red, and black spruce

Importance: Feeding by adelgid nymphs causes galls to form at the base of current-year shoots.

Abundant galls affect the appearance of trees and may even make trees unfit for sale. Individual trees can be repeatedly attacked and may accumulate dozens or even hundreds of galls.

Look For:

• *Round or pineapple-shaped, green or brown galls*, ¾ to 1 inch long, at the base of new shoots, behind the current-year growth.

Biology: Immature females overwinter near the terminal buds at the end of shoots. As buds begin to swell in spring, females mature and lay eggs. Nymphs hatch and feed at the base of the new current-year needles. This feeding causes the tree to produce galls that enclose and protect the adelgids. In late summer or early fall, galls dry out, turn brown, and open, allowing the adelgids to emerge, disperse, and reproduce.

Monitoring and Control:

• See table 1 (page 22) for degree day information.

• If galls are not abundant and are found on scattered trees, clip off and destroy green galls before they turn red and open in late summer. Old galls can be removed to improve tree appearance, but this will not provide any adelgid control.

• If galls are too numerous to hand clip, treat infested trees. Trees can be sprayed with a registered insecticide just before spruce buds break in spring or in early fall, after galls have opened. A horticultural oil or dormant oil can be used but may temporarily affect tree color.

• Some trees appear to be more resistant to this adelgid than other trees. Chip or burn severely affected trees with abundant galls.

Next Crop:

• Consider planting alternative species.

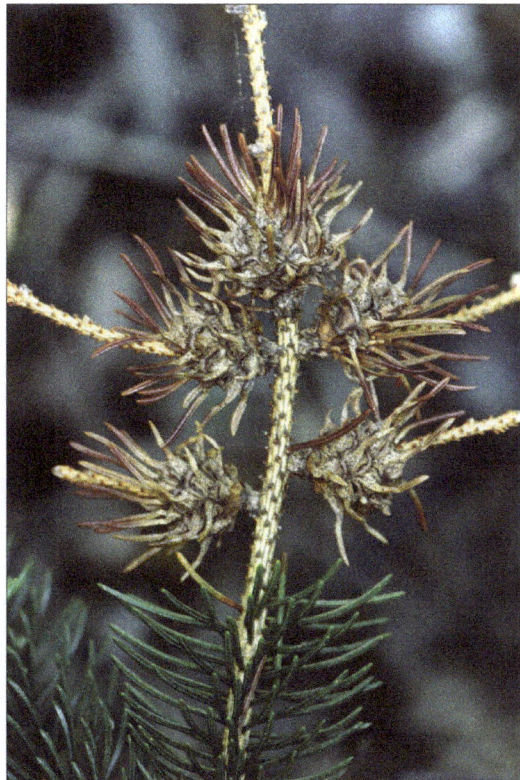

Left: Pineapple-shaped, green galls. (USFS - NCRS Archive)

Right: Older, brown-gray galls. (S. Katovich, USFS, Bugwood.org)

Northern Pitch Twig Moth (= Pitch Nodule Maker)

Retinia albicapitana

Syn. *Petrova albicapitana*

Hosts: Scotch pine; also jack pine

Importance: Caterpillars of this moth will chew away small areas of bark in the crotches of twigs and young branches. This deforms or can kill a few branch or stem tips. Damage levels are generally minor; on rare occasion this insect can degrade Scotch pine Christmas trees. This pest is usually found only within the range of jack pine, its native host.

Look For:

• *A hollow, thin-walled, brownish pitch blister* (nodule), about ½ to ¾ inch in diameter, in a branch or twig crotch.

• *Scattered flagged (discolored) branches.* Flagged branches may break off at or just beyond the blister.

• *Small, reddish-brown caterpillars*, ½ inch long, inside the blister.

Pests that cause similar symptoms: Flagged branches may be caused by eastern pine shoot borer, Diplodia shoot blight, pales weevil, pine root tip weevil, pine shoot beetle, or pine spittlebugs.

Biology: Two years are required to complete a full life cycle. A first-year larva constructs a small, blister-like nodule on a growing tip and overwinters there. The next spring, the larva moves to a twig crotch and forms a larger nodule—the one normally seen on the tree. A fully developed larva spends the second winter inside the larger nodule; pupation occurs the next spring under the nodule. The small, inconspicuous moths emerge in early summer and lay eggs, and the cycle is repeated.

Monitoring and Control: Begin looking for pitch blisters when trees reach shearing age. Treat by hand if common. Insecticides are not effective because the caterpillar is protected inside the pitch blister.

• Break open pitch blisters and crush the larvae.

• Clip off flagged, broken, or crooked branches and leaders while shearing, or simply wait for broken branch tips to fall off.

The caterpillar feeds inside a pitch blister. (USFS - NA Archive, Bugwood.org)

Spruce Gall Midge

Mayetiola piceae

Hosts: White and Black Hills spruce, occasionally Norway spruce

Importance: An occasional pest though damage requiring treatment is unusual. Individual trees can become unsalable after several years of infestation. Tiny, parasitic wasps usually keep spruce gall midges at low numbers. These small wasps can be killed by inappropriate insecticide use, which leads to an increase in gall damage.

Look For: Galls persist, so evidence of recent and past attacks can be found at any time of the year.

- *Galls or swollen twigs that tend to be elongated* and often located on only one side of a twig. These galls are not round or pineapple shaped.

- *Dead shoots* beyond the swollen twig or gall.

The adults are small flies (a midge) that can be caught in a yellow sticky trap in early spring. The larvae (immature midges) overwinter in galls on terminal shoots.

- *Bright-orange, legless midge larvae* inside galls during late summer, fall, and winter.

- *Adults that look like mosquitos* (1/16 of an inch long) but do not bite. They are dull orange to brown.

Pests that cause similar symptoms: Eastern spruce gall adelgid and other gall-forming adelgids on spruces.

Spruce gall midge pupa. (J. O'Donnell, MSU)

Orange larvae develop inside the gall. (J. O'Donnell, MSU)

Biology: Adult midges emerge from galls in late April to early May. After mating, females lay bright-orange eggs between the bud scales on new buds. The eggs hatch in 10 to 14 days, and then larvae move to the base of developing needles and begin feeding. The needles grow around the larvae, which continue to feed and develop inside small individual chambers. The bases of infested needles swell as the chambers expand to accommodate the growing larvae.

Monitoring and Control: Galls can be observed throughout the year. Note the location of any trees that show multiple galls. Hand clipping in winter or early spring may work well if populations are low. If galls are abundant on only a handful of trees, consider removing and destroying those trees. Apply insecticides with care to avoid eliminating the natural enemies that control this insect.

- Prune out and destroy galls by April 1. This technique is only useful where the trees have a few galls.

- Remove and destroy heavily infested trees.

- Use yellow sticky traps in early spring to monitor adult midge emergence.

- Time any insecticide applications to target adults and eggs, thus preventing gall formation.

Next Crop:

- If this is a serious problem in your area, consider growing other species.

Elongated galls on twigs. (S. Katovich, USFS, Bugwood.org)

Dead Tree and Stem/Root Injury

Whole or most of the tree is dead or dying. Needles fade from green to yellow to red or brown and may eventually fall off. Some trees may be leaning or fallen over. Evidence of stem injury includes masses of pitch, holes in bark, and girdling (bark removal). Your tree may be in the advanced stages of injury caused by other pests, so if you cannot find the agent here, check pests in other injury categories.

Allegheny Mound Ant

Allegheny Mound Ant

Formica exsectoides

Hosts: All Christmas tree species

Importance: Mound ants kill all vegetation within 20 feet of their mounds (nests), including seedling or sapling conifers. The ants will also kill large trees that are as far away as 50 feet if those trees shade the ant mounds.

Look For:

Groups of dead or dying trees. Search the bases of affected stems for symptoms of injury, such as small, blister-like swellings.

- *A large ant mound*, 1 to 3 feet high and 2 to 6 feet across, located among the injured trees.

- *Large ants*, ¼ inch long, with reddish-brown front ends and red tail ends. Look for ants on the mound or on live trees.

Pests that cause similar symptoms: Pocket gophers push up mounds of soil and can kill young trees in groups.

Biology: These ants kill trees that shade their mounds by injecting formic acid into the bark of the lower trunk. The ants also protect aphids and scales on neighboring live trees by chasing away the aphids' natural enemies (parasites and predators).

Monitoring and Control: Look for large ant mounds between trees or rows of trees in stands of all ages throughout the growing season. Treat individual mounds.

- Kill adult ants by applying a registered, residual insecticide to mounds any time between mid-April and October. Either level the mound with a rake before treating, or mix the insecticide into the upper 2 to 3 inches of the mound. For best results, apply just before a rain. Treat again if a new mound appears. Be careful; these ants will bite.

- Keep aphid and scale populations low.

Next Crop:

- Level and treat ant mounds before planting a new crop of trees.

Large Allegheny ant mound. (USFS - NCRS Archive)

Armillaria Root Rot (Shoestring Root Rot)

Armillaria spp.

Hosts: All trees

Importance: Species of this fungus infect roots, which slows tree growth and kills trees through girdling at the root collar. Stressed Christmas trees and trees growing on former hardwood sites near infected roots and stumps are most susceptible to infection.

Look For:
- *Yellowing and browning of all needles* on single trees or groups of trees.
- *Resin on the bark at the root collar*, where the stem and roots meet.
- *Creamy white, fanlike, leathery sheets of fungus* under the bark at the root collar.
- *Reddish-brown to black, shoestring-like fungal strands* under the bark at the root collar.

- *Clusters of honey-colored mushrooms* on or near the decaying roots or near the base of affected trees in the fall.

Pests that cause similar symptoms: Drought, pine root collar weevil, wood borers, Heterobasidion root disease, Phytophthora root rot, pine wood nematode, bark beetles

Biology: *Armillaria* can live for many years in the wood and large roots of killed trees and stumps. Reddish-brown to black fungal strands (rhizomorphs or "shoestrings") from killed trees and infected stumps grow through the soil and infect the roots of nearby trees. Vigorously growing infected trees can limit the spread of the fungus; however, once trees are under stress, they can be extensively colonized by the fungus.

Monitoring and Control: Examine the base of trees that exhibit off-color foliage for signs of the fungus.

- Remove dead trees, infected stumps, and large roots.
- Reduce tree stress by treating for other diseases, insect pests, and environmental factors. Healthy, vigorous trees are more resistant to *Armillaria* infection and damage than diseased, weak ones.

Next Crop:
- Choose a site that is well suited to the growth needs of the tree species planted. Avoid planting on sites with many large hardwood stumps.
- Remove stumps and large roots before planting.

Clockwise from top left: Armillaria mushrooms at the base of an infected tree. (F. Soukup, Bugwood.org); Dying tree near an infected stump. (J.W. Byler, USFS, Bugwood.org); White sheets of fungus (mycelial fans) under the bark at the root collar. (J. O'Brien, USFS, Bugwood.org)

Conifer Root Aphid

Prociphilus americanus

Hosts: Species of true firs (*Abies* spp.), especially Fraser fir

Importance: Conifer root aphids can be a significant pest. Ants can move aphids from large, infested fir trees to nearby interplanted seedlings that have small, susceptible root systems that are easily injured. Newly planted trees, if colonized by aphids soon after planting, are readily damaged. Fully established trees may show interior shoot stunting and discoloration due to poor root function, for which conifer root aphid can play a major contributing role.

Look For:
General decline of trees; young trees may be more heavily impacted by aphid feeding than larger, older trees that have bigger root systems.

- *Stunted young trees.*

- *Dying or dead leader and branch tips* on small trees.

- *Stunted shoot length, stunted needle length, and poor color of interior shoots* on any size fir tree.

MAY TO SEPTEMBER
- *Ant activity* at the base of fir trees.

- *Groups of waxy-appearing aphids* clustered on the roots of fir trees.

Pests that cause similar symptoms: Drought stress, excessive soil moisture, root rots, white grubs

Biology: Aphids feed by using their straw-like mouthparts to suck plant sap; in this case the aphids are piercing bark tissue covering the roots. Feeding aphids secrete honeydew, a sugar-rich plant sap that attracts ants. Ants feed on the honeydew and will protect the aphid colonies from predators. Ants will also move aphids from one tree to another. Large fir trees with big root systems are less easily damaged, whereas trees with small root systems can be impacted, especially during periods of drought. Once a colony gets established on fir, the aphids can asexually reproduce and maintain a population throughout the year. Some populations of this aphid alternate between two host plants: ash trees (spring and summer) and fir trees (other seasons). On ash, this aphid feeds on leaves.

Group of conifer root aphids. (R. Cowles, The CT Agricultural Experiment Station)

Monitoring and Control: Check trees that appear stunted; look for ants around the trunk and root collar. If you suspect root aphids, expose the roots on suspect trees, or use a shovel to sample roots at the drip line of the tree, to look for aphids. Aphid colonies on fir roots can be observed nearly any time of year. Consider treating if colonies are found. Stunting and poor foliage color in interior shoots cannot be reversed, so preventive measures are warranted when aphid colonies are found.

- Buy planting stock only from reputable nurseries; do not plant seedlings that have root aphids.

- Maintain vigorous seedlings and trees.

- If aphids are found, consider treatment.

- Apply a registered insecticide. Systemic products generally work well against aphids. However, most systemic insecticides only move upwards in trees, so products have to be thoroughly incorporated in the soil with irrigation or rainfall to reach root aphids that are feeding.

Next Crop:
- Avoid planting fir on dry, nutrient-poor sites.

- Avoid planting seedling fir among older, larger fir. Ants carry aphids from nearby older trees to seedlings that have small, susceptible root systems.

Conifer root aphids. (Richard Cowles, The CT Agricultural Experiment Station)

Stunted Fraser fir infested with root aphids; note poor color. (R.S. Kelley, VT Dept. of Forests, Parks and Recreation)

Phytophthora Root Rot

Phytophthora cinnamomi,
Phytophthora spp.

Hosts: Various species of the fungus *Phytophthora* are present throughout the United States and are known to infect several conifer species.

Importance: *Phytophthora cinnamomi* is the most important species that causes root rot of Fraser fir in the Southeastern United States. This species requires warm, wet soil and is intolerant of temperatures below freezing. Therefore, it is not likely to become established further north. Other species of *Phytophthora* can survive in cold climates and are considered a threat. Infection of roots by this fungus causes root mortality and can eventually kill trees.

Look For:
- *Foliage of seedlings or older trees changing from green to yellow to red-brown.* All foliage on the tree is likely to be affected.
- *Rotted and discolored roots.* Root tissue may be water soaked and soft. The outer layer of root tissue may be easily pulled off. Red-brown discoloration may be present in infected roots and under the bark of the root collar.
- *Stunted or wilting new growth.*

Biology: *Phytophthora* species are most common in poorly drained or heavy soils. This fungus produces spores that swim through free moisture in the soil. The fungus may survive in the soil as thick-walled resting spores or in dying plant material as thread-like hyphae.

Seedlings with low levels of Phytophthora root rot infection may leave the nursery undetected. If conditions are favorable for the fungus to develop, the disease will continue in the new plantation and the fungus may spread to healthy trees.

Monitoring and Control:
Prevention is the key to managing this disease. Conduct inspections regularly. A laboratory confirmation is recommended to ensure that correct management techniques are aimed at the specific fungus species. Take action in the nursery if Phytophthora root rot is detected. If a site becomes contaminated, disinfect tools, equipment, and muddy boots to prevent movement of the pathogen. If the site is poorly drained or has very heavy soil, consider a crop other than Christmas trees; *Phytophthora* can thrive under such conditions.

Next Crop:
- Prevent introduction of *Phytophthora* by obtaining stock from reputable sources and inspecting stock before planting.
- Do not plant seedlings or trees that show symptoms of Phytophthora root rot.

Left: Tree killed by Phytopthora root rot. (L. Haugen, USFS, Bugwood.org)

Above: Resin-soaked appearance of a Phytopthora-infected root. (E.L. Barnard, FL Dept. of Agriculture and Consumer Services, Bugwood.org)

Pine Bark Adelgid

Pineus strobi

Hosts: Eastern white pine, occasionally Scotch and Austrian pine

Importance: Pine bark adelgids weaken pine trees by sucking sap. Heavily infested trees grow poorly, become discolored, and lose their value. Some trees may die or become weak and susceptible to other pests during dry periods. However, light infestations are common and result in little damage.

Look For:

- *Discolored, stunted, weakened, or dying trees* with small but conspicuous patches of white, woolly wax on the main stem and branches. The trunk may look whitewashed.

- *Yellow or purplish insects,* less than 1/25 inch long, under the woolly wax. Use a hand lens.

MAY TO JUNE

- *Dark blue-green nymphs covered with white, waxy material* in clusters on elongating shoots.

Biology: Mature females covered with woolly wax overwinter on the tree. Eggs laid in the spring produce wingless and winged forms that infest new hosts. Immature adelgids, called crawlers, insert their mouthparts into bark and begin to feed. Populations are often controlled by natural enemies such as lady beetles and lacewings.

Monitoring and Control: Inspect trees of all ages throughout the growing season. Look for white, woolly wax and blue-green nymphs early in the growing season. Treat infested trees only if nymphs or white masses are found on numerous shoots or coat the bark. Only infested trees need to be treated. Encourage and protect natural enemy populations.

- Spray trees with a dormant oil before growth starts in the spring. Do not spray until the temperature stays above 40 °F for 24 hours. Inspect the woolly wax in early May with a hand lens to make sure the insects underneath are dead. Be sure to use adequate pressure and volume to penetrate the canopy and reach insects on the trunk.

- Alternatively, thoroughly spray trees with a registered insecticide in mid-May when the insects are active. Insecticidal soaps can be effective.

Next Crop:

- If this is a perennial problem, avoid planting eastern white pine, especially near Scotch and Austrian pine.

Heavily infested eastern white pine stem. (USFS - NA Archive, Bugwood.org)

Adelgids secrete white, waxy material as they feed. (USFS - NCRS Archive)

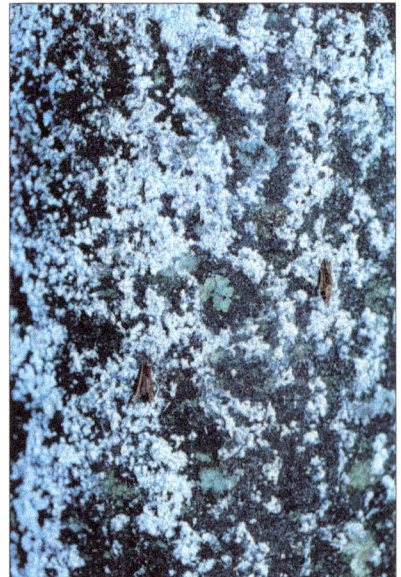

Adelgids are found under woolly wax on a trunk. (USFS - NCRS Archive)

Pine Root Collar Weevil

Hylobius radicis

Hosts: Scotch, Austrian, red, and eastern white pine

Importance: The grub-like larvae of this weevil girdle the root collar (where the stem and roots meet) and roots larger than 1 inch in diameter on young pine trees. Trees weakened by weevils may fall over and die 1 to 4 years after being attacked. This weevil is most destructive on sandy soils and where competition between trees and grass or other herbaceous vegetation is intense.

Look For:

- *Yellow to red needles on the entire tree.* Some trees, including some with green foliage, may be leaning or fallen.

- *Black, pitch-coated bark at the root collar.* Soil around the tree may be pitch soaked.

- *Yellow-white, legless, C-shaped larvae,* up to ⅓ inch long, with amber brown heads. Look for them in tunnels in the bark or in adjacent soil at or near the soil line.

LATE JUNE TO EARLY SEPTEMBER

- *White pupae,* up to ⅓ inch long, in the bark and soil where larvae are found.

Pests that cause similar symptoms: Scattered dead trees can be caused by vole feeding, Armillaria root disease, wood borers, or bark beetles. Pocket gophers can also kill trees or cause trees to lean, though they rarely damage trees greater than 1 inch in diameter.

Biology: In the spring and summer, adult weevils lay eggs at the base of pines during the day and move onto the trees at night to feed. Adult feeding causes minimal damage to twigs or shoots. Larval feeding on the inner bark of the root collar, however, injures the main stem and roots, just below the soil line. Fully-grown larvae pupate in the nearby soil. Trees respond to larval feeding by producing pitch that soaks the bark and soil. Adults emerge throughout

Black, pitch-coated bark at the root collar. (S. Katovich, USFS, Bugwood.org)

the summer, and all life stages can be found during the summer months. Because adults are weak fliers, affected trees may occur in clumps. Weevil adults will move into pine plantations from nearby infested trees.

Monitoring and Control: Begin inspecting trees when they reach 1 inch in diameter at the base. Be especially vigilant about scouting if trees are growing in sandy soils. If some trees are dying, treat all trees except healthy-looking ones that are ready for immediate harvest. If trees are not dying, look for injury (black, resin-soaked areas of bark) at the base of 20 to 30 scattered trees sometime before mid-May and again before mid-August. Treat the entire plantation if 50 percent of the inspected trees are injured.

- See table 1 (page 22) for degree day information.
- Allowing more sunlight to reach the root collar area will make conditions unfavorable for weevils. Prune off the lower whorls of branches. Removing branches growing within 1 foot of the ground will also make it easier to treat trees with an insecticide.
- Drench the root collar and a 1-foot radius of soil around each tree with a registered insecticide during warm weather to kill adults. The best time to treat is in mid-May before the adults lay eggs. Apply insecticide again in mid-August to control newly emerging weevils.
- Control weeds; pine root collar weevil damage can be concentrated along the edges of plantations where competition with herbaceous vegetation is most intense.

Next Crop:
- Plant Scotch, Austrian, or red pine at least ¼ mile away from weevil-infested pines.
- Remove any adjacent older Scotch pine plantings not being used for Christmas tree production.

Infested trees turn yellow then red-brown. (S. Katovich, USFS, Bugwood.org)

Pinewood Nematode

Pinewood Nematode

Bursaphelenchus xylophilus

Hosts: Pines, especially Scotch pine

Importance: When present in large numbers, these microscopic, parasitic roundworms can kill infested pines. The number of trees infested and killed by nematodes increases during periods of drought. Nematodes are often found in stressed trees that are dying from other causes.

Nematodes are spread by longhorned beetles. (S. Katovich, USFS)

Look For:

- *Yellowing, wilting, and browning of all needles on single or small groups of trees* during the growing season. Brown needles remain on dead trees for several months.

- *A lack of resin flow from wounds*.

Pests that cause similar symptoms: Wood borers, voles, bark beetles, and Armillaria root disease can all kill scattered trees.

Biology: Pinewood nematodes are spread from recently dead trees or logs to healthy or stressed pines in the spring by longhorned beetles (woodborers). Nematodes reproduce rapidly in the wood of infested trees during the summer, usually killing affected trees by fall.

Monitoring and Control:

Inspect plantations and closely examine any recently dead or dying Scotch pine. If a clear cause of tree death cannot be determined, consider pinewood nematode as a possible culprit.

- Have a pest specialist examine the wood of a recently killed tree to determine if nematodes are present.

- Destroy infested trees by burning or chipping before longhorned beetles emerge from them in the spring.

- Do not allow piles of cut or dead trees to build up in your plantation; longhorned beetles will develop in this material.

Next Crop:

- Avoid planting on drought-prone sites.

- Maintain healthy, vigorous trees.

All needles turn yellow to brown. (USFS - NCRS Archive)

Pocket Gopher

Geomys bursarius

Species Affected: Most Christmas tree species

Importance: Pocket gophers weaken or kill trees by feeding on their roots. Tree loss can be significant in some locations. Pocket gophers are found in the Great Plains, mainly in the western half of the North-central United States. In the Lake States, they are rarely seen east of Wisconsin. Tunneling can damage underground utilities and irrigation pipe.

Look For:

- *Ridges in the soil* caused by underground burrowing.

- *Semicircular mounds of soil.*

- *Dead trees* near mounds and ridges. These trees may be pulled easily from the ground and may have no roots remaining.

- *Destroyed tree roots.*

The pocket gopher is a burrowing rodent that is 9 to 12 inches long and has a short, sparsely haired tail. These animals have small eyes and ears; short necks; chisel-like teeth; and long, strong claws. Coat color ranges from brown to black.

Pests that cause similar symptoms: Allegheny mound ant, meadow vole, pine vole, and thirteen-lined ground squirrel can all kill young trees. Note that star-nosed moles also create ridges and mounds of soil, but do not harm trees.

Biology: Pocket gophers are solitary for much of their lives and are most active during dusk and nighttime. They are seldom seen above ground. They dig most of their tunnels in the spring and fall. They do not hibernate and prefer sandy soils. Pocket gophers are herbivores that eat whatever plant material they encounter, including tree roots.

Tree with roots chewed off by pocket gophers. (USFS - NCRS Archive, Bugwood.org)

They will pull seedlings into their tunnels.

Monitoring and Control: Inspect fields with trees of all ages in spring. Consider controlling gophers if mounds are numerous and dead trees are present.

- Limit weedy vegetation in the Christmas tree field.

- For small-scale problems, trap and hand bait pocket gophers.

- For larger populations, check your local university extension Web sites for control options. There are many treatment options available depending upon local conditions.

Semicircular mounds of soil left by pocket gophers. (USFS- NA Archive, Bugwood.org)

Rabbit

Sylvilagus floridanus

Snowshoe Hare

Lepus americanus

Species Affected: All pines, occasionally spruce and fir

Importance: Rabbits and snowshoe hares feed on the bark and the lower branches of young pines. Newly planted seedlings can be clipped off completely, especially by snowshoe hares. Snowshoe hares tend to be more damaging to conifers than rabbits. Both rabbits and hares have large population swings; when numbers are high, damage to trees can be extensive. Snowshoe hares are not as widely dispersed as rabbits and have a range that overlaps with northern conifer forests.

Look For:

- *Exposed wood* where stem and branches have been girdled or had their bark stripped. Rabbits and hares will feed as high up on the stem and branches as they can reach by standing on their hind legs.
- *Tooth marks*, 1/10 inch wide, running horizontally across the stem.
- *Smooth, clean, slanted cuts* where rabbits or hares have clipped off branches.
- *Rabbit or hare droppings and tracks* near trees.
- *In winter, well-defined runways of packed-down snow.*

Pests that cause similar symptoms: Deer, voles

Biology: Rabbits and hares may feed on tree bark during the winter, especially if populations are high and if their normal, preferred foods are scarce.

Rabbit and hare populations can vary greatly from year to year. They tend to concentrate in areas of favorable habitat, most often along brushy fencerows or around brush piles.

Monitoring and Control: Look for damage on trees of all ages during the winter when most injury occurs, especially in areas close to heavy brush. No control is needed if injury is random and infrequent. Check your local university extension Web sites for control options. There are many treatment options available depending upon local conditions.

- Discourage rabbits and hares by removing preferred cover, such as brush piles and brushy field borders.
- Hunting can be helpful in reducing rabbit or hare populations.
- Box trapping can be effective if started in late summer and continued intensively through late winter. Contact the appropriate wildlife management agency in your State for information on hunting and trapping permits.
- Encourage predators such as hawks and owls, by providing nest boxes and hunting perches.

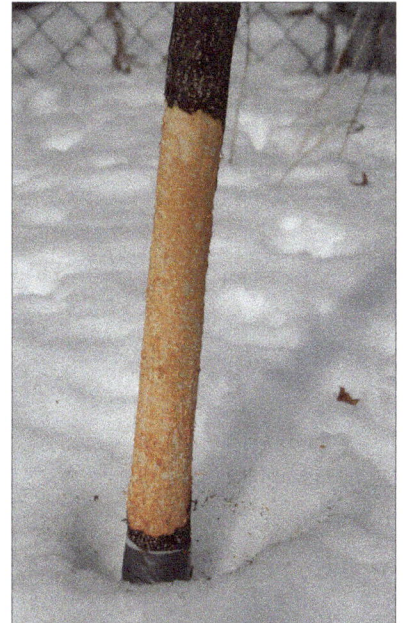

Hardwood tree girdled by rabbits. (S. Katovich, USFS, Bugwood.org)

Clean, slanted cut of twig clipped by a rabbit. (B. Marshall, Sault College, Bugwood.org)

Thirteen-Lined Ground Squirrel

Spermophilus tridecemlineatus

Species Affected: Most Christmas tree species

Importance: Thirteen-lined ground squirrels weaken or kill trees by burrowing along their roots.

Look For:

- *Burrow openings*, approximately 2 inches in diameter.
- *Semicircular mounds of soil* by burrow openings.
- *Dead trees* near mounds and burrows.
- *Damaged tree roots*.

The thirteen-lined ground squirrel is a burrowing rodent that is 5 to 8 inches long and has a sparsely haired tail. These animals have short ears, short necks, and chisel-like teeth. Coat color is brown with about 13 cream-colored stripes that are alternately solid and spotted.

Pests that cause similar symptoms: This ground squirrel is often referred to as a gopher; it should not be confused with a pocket gopher whose mounds are much larger and do not often include a well-defined burrow opening.

Biology: This ground squirrel is active by day and can be seen above ground. Christmas tree plantations with short grass cover are a favorite feeding area for ground squirrels, and the soil is usually easy for the squirrels to work. This squirrel eats weed and plant seeds and insects. The ground squirrel is a true hibernator; it spends the winter in underground burrows.

Monitoring and Control: Inspect fields in spring, especially fields with seedlings and younger trees. Consider controlling ground squirrels if burrows are numerous and young trees are dying because of injured roots.

- For small-scale problems, trap or apply baits in ground squirrel burrows.
- For larger populations, check your local university extension Web sites for control options. There are many treatment options available depending upon local conditions.
- Encourage predators such as hawks and owls, by providing nest boxes and hunting perches.

Next Crop:

- Encourage predators such as hawks and owls to hunt ground squirrels in the plantation by providing nest boxes and hunting perches.

Adult thirteen-lined ground squirrel. (S. Katovich, USFS, Bugwood.org)

Burrow opening is about 2 inches in diameter. (S. Katovich, USFS, Bugwood.org)

Meadow Vole (= Meadow Mouse)

Microtus pennsylvanicus

Pine Vole

Microtus pinetorum

Species Affected: Most Christmas tree species

Importance: Meadow voles, commonly called meadow mice, feed on bark around the base of the trunk or on lower branches. Meadow voles may kill trees by removing a complete ring of bark from the trunk (girdling). Pine voles feed on the bark of tree roots, causing short needles, yellowing of needles, and slow growth. Trees may be killed or weakened, making them vulnerable to other pests. Damage can be extensive if the right conditions exist for voles to thrive, such as heavy grass cover.

Look For:

- *Active vole runways*, 2 inches wide, devoid of live vegetation. Look for runways in areas of heavy vegetation. Runways are most visible in late winter as snow melts off fields.

- *Piles of droppings and small caches of clipped grass.*

- *Burrows 1 inch in diameter that are 2 to 4 inches deep (meadow voles) or 18 inches deep or deeper (pine voles).*

MARCH TO APRIL

- *Girdling of the trunk near the soil line*, especially on trees in heavy grass (meadow voles); *girdling of the tree's roots below the soil* (pine voles).

MAY TO JULY

- *Lack of new developing shoots, yellowing foliage, short needles.*

- *Completely brown trees* killed by girdling.

Pests that cause similar symptoms: Rabbits, hares, and porcupines can girdle trees.

Scattered dead trees can be caused by pine root collar weevil, drought, Armillaria root disease, or blister rust.

Biology: Voles are prolific breeders, and populations often increase to high numbers every 3 to 5 years. The meadow vole, commonly found in grassy

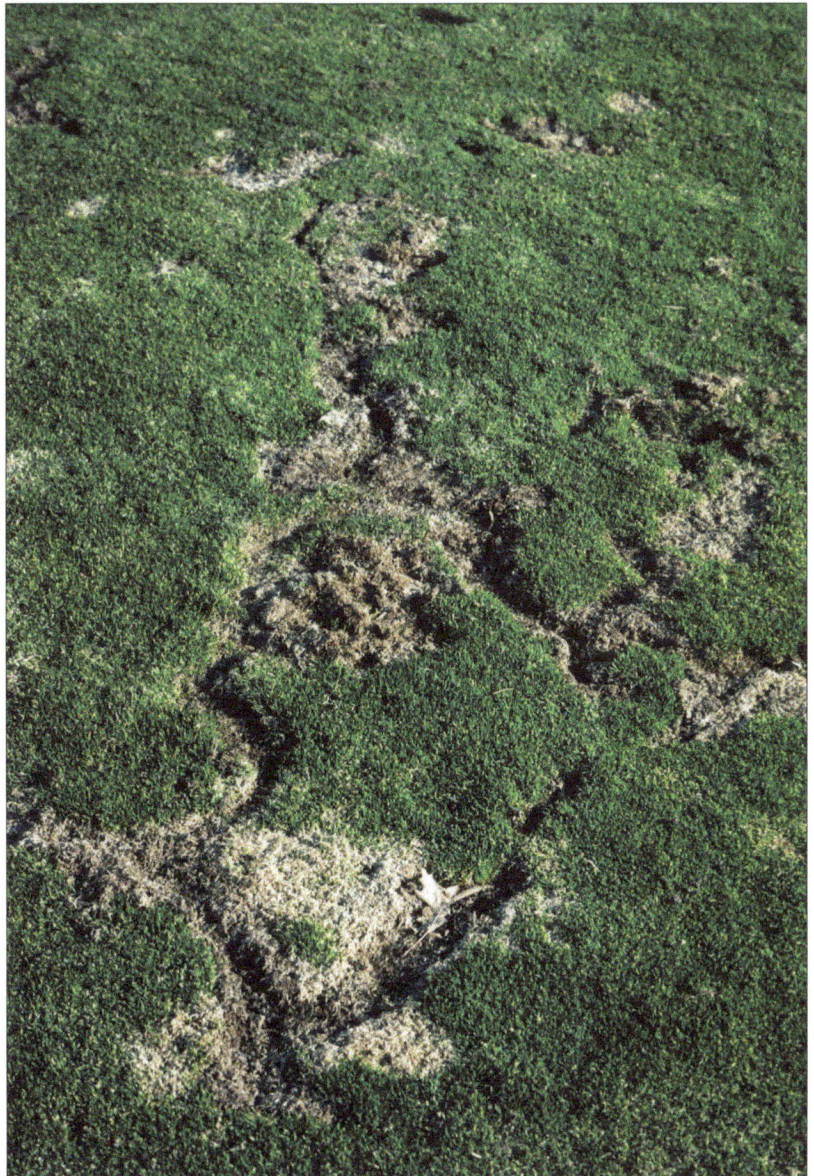

Vole runways. (USFS - NCRS Archive, Bugwood.org)

149

fields, feeds mainly on grasses and other succulent vegetation throughout the growing season. The pine vole, commonly found in shrubby, weedy fields, feeds on broad-leafed plants and their seeds. During the winter when vole populations are high and normal food supplies dwindle, voles turn to eating tree bark. Feeding generally occurs below the snow line in areas of dense, matted vegetation, so injury may not be discovered until the snow melts.

Monitoring and Control: Look for dead trees and injury in stands of all ages throughout the growing season. If you see voles and their runways or burrows frequently in the fall, consider treatment to protect trees. Voles often invade tree plantations from adjoining areas under the snow.

- Prune off girdled branches. Nothing can be done to save trees with girdled trunks.

- Mow close to the ground or apply herbicide around trees to destroy vole habitat.

- Apply a registered herbicide to eliminate vole cover.

Next Crop:

- Remove grassy vegetation on or around the site before planting and continue to control vegetation throughout the life of the trees in areas where voles are a problem.

- Encourage predators such as hawks and owls to hunt voles in the plantation by providing nest boxes and hunting perches.

Tree damaged by voles feeding on bark. (USFS - NCRS Archive, Bugwood.org)

Adult pine voles. (USFS - NCRS Archive, Bugwood.org)

White Grubs

Phyllophaga spp., *Polyphylla* spp. (May and June Beetles),

Rhizotrogus majalis (European Chafer)

Hosts: All Christmas tree species

Importance: The larvae of these beetles, called white grubs, feed on the roots of tree seedlings. The resulting damage kills seedlings or reduces their growth and vigor. Injury usually occurs during the first two growing seasons after planting and is most severe on abandoned farmland, old pastures, and grassy fields that have recently been converted to trees. White grubs are often more abundant on well-drained, sandy soils.

Look For:

- *Dead or dying seedlings* scattered throughout the stand.
- *Fibrous roots missing from dead seedlings.* Dig or gently pull up seedlings to examine roots. Damaged seedlings easily pull out from the soil; in some cases the larger roots are chewed away.

MAY TO SEPTEMBER

- *White, C-shaped larvae*, up to 1 inch long, with brownish heads and six brown legs. Most larvae will be in the upper 6 inches of soil.

Biology: White grubs normally feed on decaying organic matter, but will eat roots of grasses and tree seedlings if they encounter them. Adults are nocturnal. In May or June, the adult beetles emerge from the soil and feed on leaves of hardwood trees. Mating and egg laying follow. Eggs are laid in the soil, which is where the hatched larvae feed and do little damage the first year. Larvae of most white grub species live in the soil for 2 to 4 years, and damage to tree roots is often done by older larvae. Grassy fields and pastures can support very high populations, and mowed areas of grass may encourage egg-laying females.

Monitoring and Control:

Before Planting: The best strategy is to avoid planting seedlings into fields with dense white grub populations. Check planting sites in July of the year before planting. Run a series of furrows scattered across new planting sites and look for grubs. If you find more than 1 grub per 10 linear feet, treat seedlings just before or during planting. Another option is to eliminate grass cover and delay planting by 1 year.

After Planting: Check for white grubs monthly throughout the growing season for 3 years after planting. Treat infested blocks in the plantation if you find grub-killed seedlings.

- Apply a registered insecticide and follow directions on the label.
- Applying a fertilizer with high potassium and phosphorus and low nitrogen in the fall may stimulate root production on damaged seedlings.

White, C-shaped larvae. (S. Katovich, USFS, Bugwood.org)

Larvae eat the fine roots off seedling trees. (S. Katovich, USFS, Bugwood.org)

- Control grass competition; herbicide control is preferable over mowing during the tree seedling stage.

Next Crop:

- Use herbicides to control grasses before August in the year before planting. Do not plant seedlings into areas with heavy grass cover.

- Develop vigorous seedlings; spread roots out in the planting hole to prevent J-rooting.

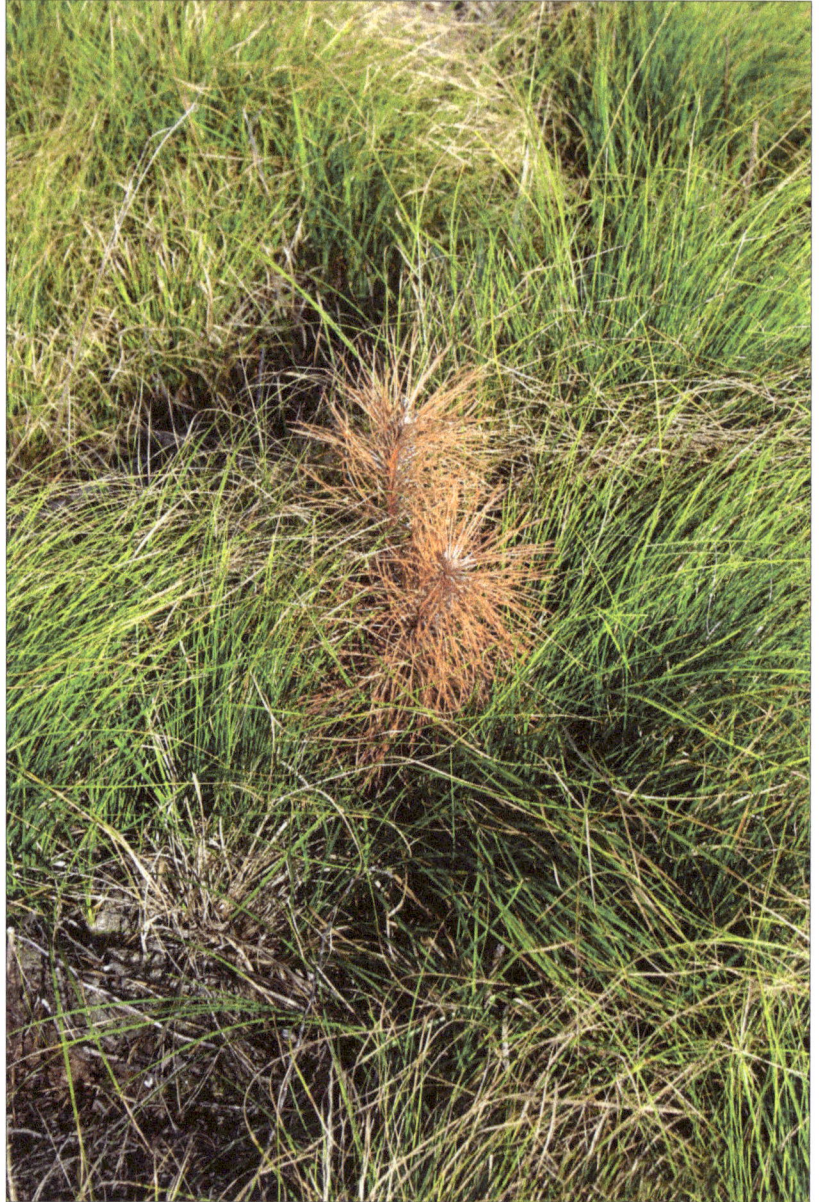

White grubs often kill seedlings growing in thick grass. (S. Katovich, USFS, Bugwood.org)

Wood Borers and Bark Beetles

Monochamus spp. (wood borers), *Ips* spp. (bark beetles); there are many bark beetle and wood borer genera.

Hosts: All Christmas tree species

Importance: The larvae of wood borers and bark beetles attack and destroy the inner bark and wood of stems and branches on Christmas trees that are severely stressed, weak, dying, or recently cut. This group of insects is generally not a concern as long as trees remain healthy.

Look For:
• *Dead or dying trees or parts of trees.*

Wood borer larva. (J. Hanson, USFS, Bugwood.org)

• *Galleries (chambers) and tunnels under the bark* that have been made by bark beetles or wood borers. These may contain white larvae that are 1/16 inch to 1 inch long, pupae, or adult beetles. Listen for borers gnawing on the wood.

• *Exit holes* that may be present on the outer bark of dead branches or trees. Exit holes left by adult bark beetles are usually abundant, round, and small (1/16 inch diameter). Adult wood borers are not as numerous as bark beetles but leave larger, oval to round holes, about the size of a pencil eraser (3/8 to 1/4 inch).

Pests that cause similar symptoms: Pine root collar weevil, eastern pine weevil, and Pales weevil all tunnel as larvae under the bark, generally in the lower portions of tree stems. Pine shoot beetles form larval galleries under the bark.

Biology: Adult wood borers and bark beetles lay eggs in or just under the bark of trees that are weak, dying, or recently killed or felled. Bark beetle and wood borer larvae feed in tunnels in the inner bark. Large wood borer larvae will also tunnel into the wood. Trees may be colonized by several different woodborer and bark beetle species. Bark beetles may emerge from one tree and attack surrounding trees.

Monitoring and Control: Inspect trees of all ages during and after periods of drought, ground fires, and other disturbances. Look for pockets of dead trees and treat accordingly.

• See table 1 (page 22) for degree day information under pine engraver.

• Remove and burn or chip dead or infested trees.

• Do not leave large piles of cut or discarded trees in your

Wood borer exit hole and fibrous frass. (USFS - NCRS Archive)

plantations during the spring and summer months.

- Look for the primary source of the problem to determine what is weakening or killing trees before the borers or beetles infest them.

- Maintain vigorous trees; irrigate as needed.

Next Crop:

- Do not replant on stressful or marginal sites until you have identified and controlled the problem that weakened the trees.

Ips bark beetle exit holes. (S. Katovich, USFS, Bugwood.org)

Ips bark beetle gallery. (S. Katovich, USFS, Bugwood.org)

Yellow-Bellied Sapsucker

Sphyrapicus varius

Species Affected: Scotch and Austrian pine, occasionally other conifers

Importance: Sapsuckers are members of the woodpecker family that peck holes in the bark of larger Christmas trees, causing sap flow on branches and stems. They often repeatedly visit individual trees because of the quantity or sugar content of the sap. When severely injured by feeding sapsuckers, branches and trees may be girdled and die. The feeding injury also permits insects and pathogens to enter through the bark.

Look For:

- *Evenly spaced rows of large pits or holes ¼ inch or larger on the trunk of the tree.*

- *A robin-sized woodpecker* with bands of red, black, and white on the head. The belly is yellow and the back is white and black. The male has a red throat; the female has a white throat.

Biology: The sapsucker is a forest bird that feeds on tree sap and the insects that get caught in the sap flowing from the wounds on branches and trunks. This migratory bird is a common summer resident in the Northern States, but rarely a winter resident.

Monitoring and Control: The yellow-bellied sapsucker is protected by the Federal Migratory Bird Treaty Act. Keep in mind that control is difficult and may not be justified in most cases.

- Leave trees being repeatedly injured. Sapsuckers will continue to feed on these trees, which will protect adjacent trees.

Sapsucker feeding holes on stem of a white pine. (S. Katovich, USFS, Bugwood.org)

Adult sapsucker. (J.N. Dell, Bugwood.org)

155

Zimmerman Pine Moth

Dioryctria zimmermani

Hosts: All pines, especially Scotch and Austrian; rarely spruces and firs

Importance: Caterpillars feed just under the bark on the trunk, large branches, or occasionally on the terminal leader. Pitch masses on the stem may affect the appearance of trees and reduce their value. Feeding may result in dead shoots or a dead leader, or weak or broken branches. Repeated attacks on the trunk may cause trees to break off at the site of the injury.

Look For:

- *Coagulated pitch masses* on the trunk, often at a branch whorl. Pitch masses may also be found on the underside of branches or on shoots near the terminal leader. Reddish frass may be mixed in with the pitch.

- *A trunk that may be swollen above the mass, or may break* if the tree is heavily injured. On Scotch pine, the attack site may be on a gall caused by one of the gall rusts of pines.

- *Branches that are dead or broken* off from the trunk.

- *A discolored or broken leader* (sometimes a lateral shoot) directly above a mass of coagulated, white or pinkish pitch.

APRIL TO JULY

- *Pinkish-green larvae*, up to ¾ inch long, in tunneled areas under soft pitch masses.

JULY TO AUGUST

- Brown pupae or empty pupal cases, ¾ inch long, at the exit of feeding tunnels from mid-July to late August.

Biology: This native insect has one generation per year. Tiny caterpillars overwinter in bark cracks and crevices. They become active in spring, between early April and early May, when they bore under the bark and feed in tunnels in the phloem (inner bark) for several weeks. Pitch emerges from the wound site and the feeding larvae push reddish frass out of the wound during the summer, creating the characteristic pitch mass. Larvae pupate near the entrance to their feeding tunnel. Adult moths emerge between mid-July and late August, mate, and lay eggs on the bark. Upon hatching, the caterpillars spin tiny silk cases on the bark where they overwinter.

Dead, collapsed branch damaged at the stem. (S. Katovich, USFS)

Monitoring and Control: Begin scouting fields for damage when trees are 4 to 5 years old. On Scotch and Austrian pine, look for pitch masses on the main stem or the underside of large branches. On Austrian and other pines, look for pitch masses on shoots near the terminal leader.

- See table 1 (page 22) for degree day information.

- Some individual trees act like "brood trees" and are repeatedly attacked. Female moths may be attracted to fresh pitch masses from earlier attacks and lay eggs on those trees year after year. Chip or burn heavily infested trees by early July to destroy the developing insects.

- If infestations are light, pitch masses can be cut out using a pocket knife or shearing tool.

- Hand prune and destroy occasional injured shoots.

- When damage is widespread, apply a persistent registered insecticide between early April and early May as the weather warms. Larvae are vulnerable to insecticides in spring before they bore under the bark. Use enough nozzle pressure and water to drench bark on the trunk and branches.

Large pitch masses on the stem are often near a branch whorl. (D. McCullough, MSU)

General Index

A

Acantholyda erythrocephala ...75

Adelges abietis 131

Adelges cooleyi 128

Adelges piceae 91

adelgids

    balsam woolly 91

    cooley spruce gall............. 128

    eastern spruce gall 131

    pine bark 142

Admes mite 28

air pollution injury 29

Allegheny mound ant............. 137

Anomala beetle.......................74

Anomala oblivia......................74

ant, Allegheny mound........... 137

aphids

    balsam twig........................32

    conifer root...................... 139

    spotted pine.......................89

    white pine...........................89

Aphrophora parallela 113

Aphrophora saratogensis 115

Argyrotaenia pinatubana 77

Armillaria root rot 138

Armillaria spp. 138

B

bagworm 65

balsam fir sawfly 66

balsam gall midge...................30

balsam shootboring sawfly 90

balsam twig aphid....................32

balsam woolly adelgid 91

bark beetle 153

    Ips........................... 153

    pine shoot 111

beetle

    Anomala 74

    bark 153

    jack pine tip...................... 103

    pine shoot 111

borer, eastern pine shoot........98

borers, wood.......................... 153

broom rust of fir 93

brown spot needle blight.........34

budworm

    jack pine.............................73

    spruce 82

Bursaphelenchus
    xylophilus 145

C

canker

    Cytospora (see Leucostoma)

    Diplodia 96

    Gremmeniella
    (see Scleroderris)

    Leucostoma 104

    Phomopsis 108

    Scleroderris...................... 117

    Sphaeropsis (see Diplodia)

cedar-apple rust 127

chafer, pine 74

Chionaspis pinifoliae 50

Choristoneura fumiferana........82

Choristoneura pinus pinus 73

Chrysomyxa spp. 58

Cinara spp..............................89

Coleosporium asterum 48

Coleotechnites piceaella 84

conifer root aphid.................. 139

Conophthorus resinosae 103

Contarinia baeri 76

Contarinia pseudotsuga 37

Cooley spruce gall adelgid 128

cottontail rabbit..................... 147

Cronartium quercuum 130

Cronartium ribicola 120

Cyclaneusma minus................35

Cyclaneusma needlecast 35

Cytospora canker
    (see Leucostoma)

D

Dasychira pinicola...................78

Dasychira plagiata 78

deer 94

Delphinella balsameae 95

Delphinella shoot blight 95

Dichomeris marginella 6

Dioryctria zimmermani...........156

Diplodia pinea 96

Diplodia shoot blight and
    canker................................96

Diprion similis..........................72

Dothistroma needle blight 36

Douglas-fir needle midge 37

drought injury..........................38

E

eastern gall rust 130

eastern pine shoot borer 98

eastern pine weevil 100

eastern spruce gall adelgid ... 131

elongate hemlock scale...........39

Endocronartium harknessii....130

Epinotia nanana 84

eriophyid mites 40

Eucosma gloriola 98

Eulachnus agilis 89

European chafer 151

European pine sawfly 67

European pine shoot moth 101

Eurytetranychus admes...........28

F

fall needle drop 41

Fiorinia externa 39

fir needle rust 42

Formica exsectoides............. 137

frost injury 102

G

gall rusts............................... 130

Geomys bursarius................. 146

Gnophothrips spp....................51

gopher, pocket...................... 146

grasshoppers...........................69

green spruce needleminer.........5

Gremmeniella abietina........... 117

Gremmeniella balsamea........ 117

Gremmeniella canker (see
    Scleroderris canker)

grosbeak, pine 109

ground squirrel,
    thirteen-lined.................... 148

grubs, white 151

Gymnosporangium
    juniperi-virginianae 127

gypsy moth 70

H

hare 147

herbicide injury 43

Hylobius pales.......................106

Hylobius radicis 143

Hylobius rhizophagus 110

I

introduced pine sawfly 72

Ips spp. 153

Isthmiella faullii.......................44

Isthmiella needlecast 44

J

jack pine budworm 73

jack pine tip beetle................ 103

L

Lepus americanus................. 147

Leucostoma canker 104

Leucostoma kunzei 104

Lirula mirabilis 44

Lirula nervata 44

Lirula needlecast.....................44

Lophodermium needlecast 46

Lophodermium seditiosum......46

Lymantria dispar 70

M

Macaria signaria......................83

Mayetiola piceae 133

meadow vole......................... 149

Melampsorella caryophyllacearum 93

Melanoplus spp. 69

Microtus pennsylvanicus 149

Microtus pinetorum 149

midge

    balsam gall 30

    douglas-fir needle...............37

    pine needle..........................76

    spruce gall.........................133

Milesina spp. 42

Mindarus abietinus 32

mites

    admes.................................28

    Eriophyid 40

    spruce spider 59

Monochamus spp. 153

moth

    European pine shoot........101

    gypsy.................................70

    Nantucket pine tip.............105

    northern conifer tussock moth 78

    northern pitch twig...........132

    pine tube 77

    pine tussock 78

    Zimmerman pine 156

mouse (see meadow vole)

Mycosphaerella dearnessii 34

Mycosphaerella pini 36

N

Naemacyclus needlecast (see Cyclaneusma needlecast)

Nalepella spp. 40

Nantucket pine tip moth 105

needle blight

    brown spot 34

    Dothistroma.......................36

    Rhizosphaera 54

needle midge

    Douglas-fir..........................37

    pine....................................76

needle rust

    fir.......................................42

    pine....................................48

    spruce 58

needlecast

    Cyclaneusma 35

    Isthmiella 44

    Lirula 44

    Lophodermium....................46

    Naemacyclus.....................35

    Ploioderma.........................52

    Rhabdocline.......................53

    Rhizosphaera 55

    Swiss 60

Needleminer, spruce 84

nematode, pinewood...........145

Neodiprion abietis...................66

Neodiprion lecontei 80

Neodiprion sertifer..................67

northern conifer tussock moth 78

northern pitch twig moth 132

O

Odocoileus virginianus 94

Oligonychus ununguis 59

P

Pales weevil 106

Paradiplosis tumifex 30

Peridermium harknessii 130

Petrova albicapitana 132

Phaeocryptopus gäeumannii 60

Phomopsis canker 108

Phomopsis spp. 108

Phyllophaga spp. 151

Physokermes piceae 119

Phytophthora cinnamomi 141

Phytophthora root rot 141

Phytophthora spp. 141

pine bark adelgid 142

pine chafer 74

pine false webworm.................75

pine grosbeak 109

pine needle midge 76

pine needle rust 48

pine needle scale....................50

pine root collar weevil............143

pine root tip weevil................110

pine shoot beetle 111

pine spittlebug 113

pine thrips 51

pine tortoise scale 114

pine tube moth 77

pine tussock moth 78

pine vole 149

pine webworm.........................79

Pineus strobi 142

pinewood nematode............145

Pinicola enucleator 109

Pissodes nemorensis............100

Pissodes strobi 122

pitch nodule maker 132

Pleroneura brunneicornis 90

Ploioderma lethale 52

Ploioderma needlecast...........52

pocket gopher......................146

Pococera robustella...............79

Polyphylla spp.......................151

Prociphilus americanus 139

R

rabbit....................................147

redheaded pine sawfly 80

Retinia albicapitana 132

Rhabdocline needlecast.........53

Rhabdocline pseudotsugae 53

Rhizosphaera kalkhoffii............55

Rhizosphaera needle blight of firs 54

Rhizosphaera needlecast of spruce 55

Rhizosphaera pini 54

Rhizotrogus majalis...............151

Rhyacionia buoliana..............101

Rhyacionia frustrana 105

root rot

    Armillaria (shoestring) 138

    Phytophthora 141

rust

    broom of fir.........................93

    cedar-apple.......................127

    eastern gall.......................130

    fir needle...........................42

    gall...................................130

    pine needle.........................48

    spruce needle 58

    western gall 130

    white pine blister 120

S

salt injury.................................57

sapsucker, yellow-bellied 155

Saratoga spittlebug 115

sawfly

    balsam fir............................66

    balsam shootboring 90

    European pine 67

    introduced pine 72

    redheaded pine 80

scale

    elongate hemlock...............39

    pine needle.........................50

    pine tortoise 114

    spruce bud 119

Scleroderris canker...............117

Setomelanomma holmii..........55

Setoptus spp.........................40

shoot blight

    Delphinella..........................95

    Diplodia 96

    Sirococcus 118

    Sphaeropsis (see Diplodia)

Sirococcus conigenus 118

Sirococcus shoot blight.........118

snowshoe hare......................147

Spermophilus tridecemlineatus 148

Sphaeropsis shoot blight and
   canker (see Diplodia)

Sphyrapicus varius.................155

spider mites (see spruce spider
   mite)

spittlebug

   pine....................................113

   Saratoga............................115

spotted pine aphid...................89

spruce bud scale 119

spruce budworm.....................82

spruce fir looper.....................83

spruce gall midge 133

spruce needle rusts 58

spruce needleminers 84

spruce spider mite..................59

Stigmina lautii 55

Stigmina needlecast 55

Sudden needle drop 55

Swiss needlecast.....................60

Sylvilagus floridanus 147

T

Taniva albolineana 84

thirteen-lined
   ground squirrel..................148

thrips, pine 51

*Thyridopteryx
ephemeraeformis*...............65

Tomicus piniperda 111

Toumeyella parvicornis 114

U

Uredinopsis spp.......................42

V

voles

   meadow.............................149

   pine...................................149

W

webworm

   pine.....................................79

   pine false.............................75

weevil

   eastern pine 100

   Pales.................................106

   pine root collar 143

   pine root tip 110

   white pine.........................122

Weir's spruce cushion rust 58

western gall rust....................130

white grubs 151

white pine aphid......................89

white pine blister rust 120

white pine weevil...................122

winter injury............................61

wood borers...........................153

Y

yellow-bellied sapsucker.......155

Z

Zimmerman pine moth 156

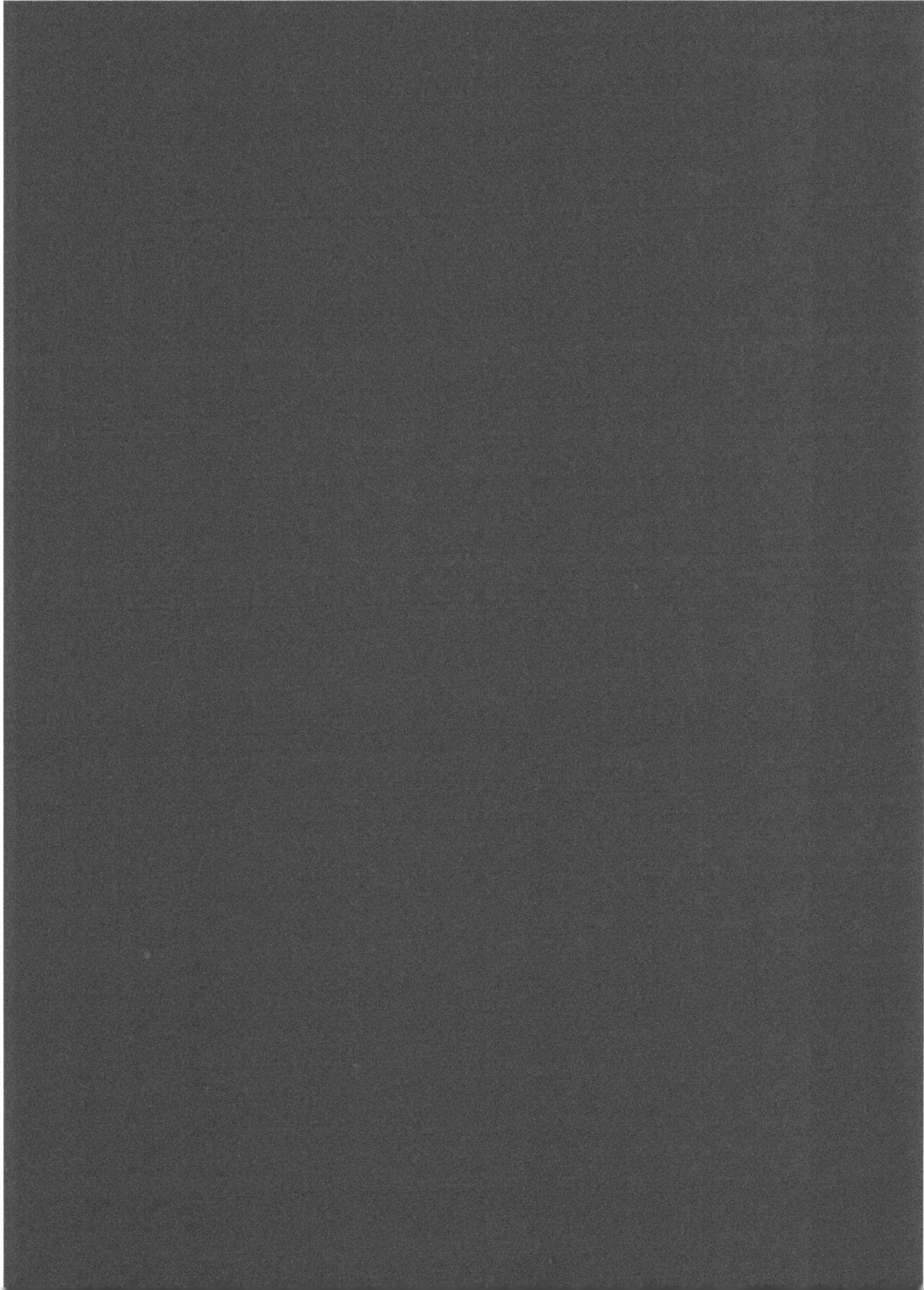

www.ingramcontent.com/pod-product-compliance
Lightning Source LLC
Chambersburg PA
CBHW041431270326
41935CB00020B/1841